Climate Change and Trade

Climate Change and Trade

Taxing carbon at the border?

Daniel Gros

and

Christian Egenhofer

in collaboration with

Noriko Fujiwara

Selen Sarisoy Guerin

and

Anton Georgiev

Centre for European Policy Studies

Brussels

The Centre for European Policy Studies (CEPS) is an independent policy research institute based in Brussels. Its mission is to produce sound analytical research leading to constructive solutions to the challenges facing Europe today. CEPS Paperbacks present analysis and views by leading experts on important questions in the arena of European public policy, written in a style geared to an informed but generalist readership.

This report was commissioned by Confindustria, the Italian employers' federation. The research was carried out by a team of highly experienced CEPS researchers: Daniel Gros, Director and Head of Economic Policies; Christian Egenhofer, Senior Research Fellow; Noriko Fujiwara, Research Fellow and Head of Climate Change; Selen Sarisoy Guerin, Associate Research Fellow, Head of Trade Policy and Lecturer in International Trade at Vrije Universiteit Brussel; and Anton Georgiev, Researcher. The views expressed in this report are those of the authors writing in a personal capacity and do not necessarily reflect those of CEPS or Confindustria or any other institution with which the authors are associated.

ISBN 978-92-9079-867-5

Centre for European Policy Studies
Place du Congrès 1, B-1000 Brussels
Tel: 32 (0) 2 229.39.11 Fax: 32 (0) 2 219.41.51
e-mail: info@ceps.eu
internet: http://www.ceps.eu

CONTENTS

Executive Summary 1
1. Introduction 7
   1.1 The EU as the first mover 8
   1.2 ... to be reproduced in the US? 9
   1.3 But what about the rest of the world? 10
   1.4 Tariffs are always bad? 13
   1.5 The physics and politics of different fossil fuels: Hydrocarbons versus pure carbon (coal) 13
2. EU Climate Change Strategies for the post-2012 period 15
   2.1 The strategic approach 15
   2.2 The climate and energy package 17
   2.3 Implementation 18
   2.4 Costs of the integrated package 21
   2.5 The EU ETS and the development of international carbon markets 22
   2.6 The EU ETS in the third phase (2013-20) 23
   2.7 Concluding remarks 25
3. Global Welfare Implications of Carbon Border Taxes 26
   3.1 Introduction 26
   3.2 A simple illustration of the welfare gain from the introduction of a carbon tariff 29
   3.3 In the absence of a carbon tariff, a cap-and-trade system can be counterproductive 31
   3.4 Model based analysis confirms these points 37
   3.5 Concluding remarks 38
4. Compatibility of Carbon Border Measures with WTO Rules 39
   4.1 BTAs in the WTO law: A review of the literature 43
   4.2 The relevant WTO regime for a EU carbon border tax 45
      4.2.1 Establishing the carbon content 49
      4.2.2 Article XX 56

4.3 Case law ... 59
4.4 Key elements for the design of a WTO-compatible EU carbon tax ... 61
4.5 Conclusion ... 62
5. The Criteria of the UN Framework Convention for Climate Change ... 66
5.1 Key indicators for climate change ... 67
5.1.1 The stock of historical emissions ... 69
5.1.2 Per capita emissions ... 69
5.1.3 CO_2 intensity of the economy ... 70
5.2 Mid- and long-term global emissions targets ... 75
5.3 Methodologies for allocation of responsibilities ... 77
5.4 Methodologies and data for comparability of efforts ... 80
5.5 Concluding remarks ... 84
6. Enforcement and Compliance ... 86
6.1 The Kyoto Protocol in the first commitment period (2008-12) ... 87
6.1.1 Commitments, achievements and prospects for compliance ... 87
6.1.2 How does the Kyoto Protocol compliance system work? ... 91
6.2 Post-2012 discussions ... 93
6.2.1 Ensuring enforcement of pledged domestic actions by developing countries ... 93
6.3 UNFCCC and the Kyoto Protocol on trade measures ... 94
6.4 Summary and concluding remarks ... 96
7. Conclusions ... 98
References ... 102

List of Tables

Table 2.1 National overall targets for the share of energy from renewable sources in gross final consumption of energy in 2020 and GHG emissions limits in non-ETS sectors for the period 2013-20 20
Table 3.1 Carbon intensity of exports and GDP in selected key economies.......34
Table 4.1 Top five exports in US-China trade, ranked by their CO_2 embodiment in 2003 ($MtCO_2$) 50
Table 4.2 CO_2 emissions[a] by sector, EU-27 (shares, %)...58
Table 5.1 Key statistics on selected Annex I parties...74
Table 5.2 Key statistics on carbon emissions in emerging economies................75
Table 5.3 Overview of commitments by regions for different target scenarios.76
Table 5.4 Global shares of population, income, capacity, cumulative emissions, responsibility and obligation (RCI) for selected countries and groups of countries (%)..79
Table 6.1 Key statistics of selected Annex I parties...88

List of Figures

Figure 1.1 Share of different anthropogenic GHGs in total world emissions in 2004 in terms of carbon dioxide equivalents (CO_2-eq) 8
Figure 3.1 Equilibrium without tariff 29
Figure 3.2 Equilibrium with tariff 30
Figure 3.3 Effect of ETS with higher carbon intensity abroad.............................35
Figure 5.1 Annual per capita emissions, by country, 2005 (tCO_2e) 70
Figure 5.2 CO_2 intensity trends in selected economies, 1990-2006 (percentage growth from 1990) 72
Figure 6.1 Canada's GHG emissions 1990-2007 90

List of Boxes

Box 1.1 The role of CO_2..7
Box 4.1 Article XX GATT/WTO: General exceptions...40
Box 4.2 Standardisation at ISO 51
Box 4.3 Companies measuring GHG emissions in their supply chain 52
Box 4.4 Benchmarking for the EU ETS 53
Box 4.5 Bottom-up approaches to carbon footprint labelling by Tesco 54
Box 4.6 Superfund chemical excises...64

EXECUTIVE SUMMARY

The EU has been a pioneer in global efforts to combat climate change through its decision to use a cap-and-trade system, with the aim of reducing greenhouse gas (GHG) emissions by 20% by 2020. This target could reach as high as 30% depending on other countries' commitments. The EU emissions trading system (EU ETS) in its current form is already imposing costs on industry in the EU and these costs can be expected to increase under the post-2012 regime that the EU has in principle already decided upon. Other developed countries such as the US are widely expected to introduce a cap-and-trade system as well, but most emerging economies have no intention to follow any time soon.

The UN Framework Convention on Climate Change (UNFCCC) states that developed and developing countries have "common but differentiated responsibilities and respective capabilities". There is a consensus that developed countries must reduce their emissions first - reflecting both the principles of historical responsibilities and capabilities - while at the same time developing countries need to put into place measures to curb their emissions, yet falling short of introducing economy-wide, legally binding commitments, such as an emissions cap. Developing countries are wary that an economy-wide cap would undermine economic growth, for example, by restricting the use of coal, which in many cases is domestically available. The recent offer from the Chinese government to reduce the emissions intensity of its economy by 40-45%,[1] but rejecting any

[1] The official letter to the UNFCCC Secretariat, states: "China will endeavour to lower its carbon dioxide emissions per unit of GDP by 40-45% by 2020 compared to the 2005 level" (see "Appendix II – Nationally appropriate mitigation actions of developing country Parties", available at http://unfccc.int).

overall ceiling, suggests the limits of what is acceptable to major developing countries.

A global cap-and-trade system encompassing all major emitters is thus at this stage not possible. Hence, large differences in the price of carbon, both explicitly and implicitly, are likely to persist.

From a purely economic perspective, a straightforward way to move towards a global, 'level' pricing of carbon would be for the EU to impose an import tax on the content of CO_2 of all goods imported into the EU from countries that do not have their own cap-and-trade system or equivalent measures. The main argument for such a move is that such a 'carbon' import tax would establish a 'shadow' carbon price even in the rest of the world.

This study analyses the economic and political consequences of such a tax and whether it would be compatible with WTO rules. The major findings are:

1. A CO_2 border tax or import tariff would increase global welfare.
2. Such a carbon import tariff can be made to be compatible with WTO rules.
3. There are no insurmountable practical obstacles to introducing such a tariff.
4. The equity concerns of the UNFCCC could be taken into account by rebating the proceeds of the tariff to those countries manifestly unable to shoulder the burden themselves.

These four points are linked and require some background, which is presented below.

1) Justification for a carbon import tariff

This is the fundamental point of this study in many respects. Simple modelling shows that a carbon import tariff is a useful complement to a domestic ceiling on emissions, as provided for in the EU ETS. The intuition behind this general result is clear: an import tariff improves global welfare because it transfers, at least partially, via trade flows, carbon pricing even to those parts of the world where governments have so far refrained from imposing domestic measures of any magnitude. In other words, it creates a mechanism that enforces the pass-through of carbon costs across the globe, therefore making domestic consumers pay the full cost of carbon. A key effect of such a tariff is that it would *always* lower global emissions. This is a very general result, which does not depend on what specific model one has in mind since a carbon import tariff would reduce EU imports of energy-intensive goods, thus reducing emissions abroad. Since the ETS provides a ceiling on emissions in Europe, it follows immediately that a carbon import tariff will lead to a fall in global emissions.

By contrast, a 'stand-alone' EU ETS risks being *ineffectual* because the ETS will lead to higher production of energy-intensive goods and thus higher emissions in countries without a carbon price (resulting in so-called 'carbon leakage'). The available evidence on the importance of carbon leakage is sketchy. Studies focusing on the limited number of energy-intensive industries have generally found a low potential for carbon leakage, but this is due to the importance of sectors whose output is not traded intensively, such as electricity and cement. However, the potential for carbon leakage increases considerably if one takes into account the fact that all products from the sectors covered by the ETS are important inputs throughout the economy (i.e. counting the embedded carbon). Studies that take these indirect channels for carbon leakage into account arrive at much higher estimates.

The potential for carbon leakage is a key unresolved empirical issue because, as shown below, it is possible that a 'stand-alone' ETS is not only ineffectual, but actually leads to higher global emissions if production abroad is more carbon-intensive than in the EU.

Another way to transfer the price signal on carbon to the rest of the world would be via the generalised use of the so-called 'Flexible Mechanisms', such as the Clean Development Mechanism (CDM) under which credits valid under the domestic 'cap-and-trade' system can be earned from projects in developing countries that reduce emissions below baselines. However, as developed below, the CDM, while useful in many instances, cannot on its own establish a global shadow price for carbon given the lack of scale and problems with host-country control.

2) WTO-compatibility

The preceding result is a key condition to make border measures compatible with the basic rules of the World Trade Organisation (WTO). In general the WTO rules are very restrictive on any border measure, but Article XX (g) provides for a general exemption for measures "relating to the conservation of exhaustible natural resources if such measures are made effective in conjunction with restrictions on domestic production". In the specific case of a border tax on the CO_2 content of an imported good, the EU can argue that this benefits the atmosphere and that the EU already has domestic restrictions on domestic production. This double requirement cannot of course be invoked by other countries that do not have a domestic carbon price and hence should not pose a danger of triggering a generalised trade war.

Other conditions for WTO-compatibility are:

- The tariff rate on any product should not be higher than its carbon content times the difference between the carbon price in the EU and abroad. Products from the US would thus not be taxed if the US introduces its own cap-and-trade system with a similar target – leading to a similar carbon price – to that adopted in Europe.
- The carbon tax should be revenue-neutral: revenues collected from the carbon tax should be used to create a fund to finance the transition of energy-intensive industries and to invest in new technologies for climate change not only in the EU but globally and distributed according to the UNFCCC criteria of 'responsibility' and 'capability'.

3) Practical implementation

Imposing an import tariff on the CO_2 content of imports is widely considered unrealistic because it would be difficult in practice to measure the CO_2, or carbon footprint of all products. However, the exercise of calculating a product's carbon footprint is becoming more and more common, and an objective norm is already developing in the form of ISO 14067, which can serve as the 'external' benchmark required by WTO rules.

Pressure from the market where consumers want to be informed about the carbon footprint of the products they buy has already led to initiatives by major multinational retailers to provide this information. Since a large part of consumption goods are imported from major emerging markets, this implies that producers in China, for example, will in many cases have to publish the carbon footprint of their products in order to get them on the shelves in supermarkets in the EU. As this trend will continue, the practical problems of implementing a carbon-based border measure should diminish over time and should not be considered an insurmountable obstacle.

4) Differentiated responsibilities and comparability of efforts remain important issues

The UNFCCC lays down the principle of "common but differentiated responsibilities and respective capabilities". This means that developed countries ('Annex 1' under the Kyoto Protocol) have more responsibility than developing countries ('Non-Annex 1' under the Kyoto Protocol) and should take the lead in climate action. That much is generally agreed. However, with China overtaking both the US and the EU as a source of CO_2 emissions and with developing countries projected to be responsible for some three-quarters of primary energy demand growth by 2030, global climate change targets can only be met if developing countries start implementing strong climate policies now.

So far developing countries have been willing only to consider bottom-up approaches based on domestic policies and measures reflecting their own national circumstances and priorities. As long as these countries are not willing to discuss common indicators for comparability of efforts with developed countries, or among themselves, it is doubtful that the bottom-up approach will lead to meeting global climate change targets. A second-best tool could be found in a shadow carbon price set through border measures.

However, the equity concerns implicit in the 'differentiated responsibilities' should be addressed. The most straightforward way to do so would be to rebate the proceeds of the import tariff according to the UNFCCC criteria of "responsibility' and 'capability'. Both legal considerations under the WTO and the equity concerns expressed by the UNFCCC would thus point in the same direction. The equity argument is simply that for the poorest countries a domestic cap-and-trade system (or carbon tax) would have unacceptably negative consequences for growth. In this case the proceeds collected by the EU at the border could be spent in these countries on further mitigation efforts. These rebates should be additional to any funding to be agreed anyway in the global negotiations.

More in general, the UNFCCC also raises the issue of the comparability of efforts. This has a number of dimensions. It is widely estimated that developed countries will need to spend about 1% of their GDP on energy savings and other mitigation efforts. What level of effort as a share of GDP should be considered 'equivalent' for a developing country? Moreover, how can one compare a cap-and-trade system to a commitment to invest huge sums in renewable energy? Should one compare the expenditure or the impact in terms of emissions avoided?

These four key issues, inter alia, are discussed in detail in this report. It is organised along the following lines:

- Chapter 1 provides a general introduction to the overall issue.
- Chapter 2 provides an overview of EU strategy for climate policy beyond 2012.
- Chapter 3 contains a detailed discussion of the impact of border measures on global welfare on the basis of a general theoretical model.
- Chapter 4 then discusses, on the basis of the results of chapter 2, under what conditions a border tax on the CO_2 content of imports would be compatible with WTO norms.
- Chapter 5 attempts to find out the extent to which commitments already taken or about to be adopted elsewhere might be comparable to what is planned for the EU.
- Chapter 6 describes the weaknesses of the enforcement mechanism of the Kyoto Protocol.
- Chapter 7 concludes with some general considerations.

1. INTRODUCTION

Combating climate change (or rather preventing an excessive warming of the earth) is now a key policy issue in most parts of the world. A series of major international scientific reports has highlighted the likely costs resulting from the increase in global temperatures that will follow from an unrestrained increase in the emissions of greenhouse gases (GHGs). Governments have taken notice, and, especially in Europe, they are increasingly willing to take action to combat emissions of the most important greenhouse gas, namely CO_2 (Box 1.1), which this report uses as a proxy for all GHGs.

Box 1.1 The role of CO_2

That GHG emissions are causing global warming is now well established. In its latest assessment – Fourth Assessment Report (AR4) – the Intergovernmental Panel on Climate Change (IPCC, 2007) observed that the warming of the climate system is "unequivocal" and that "most of the observed increase in global average temperatures since the mid-20th century is *very likely* due to the observed increase in anthropogenic greenhouse gas concentrations." Hence, reducing GHG emissions would help to address "dangerous climate change".

The Kyoto Protocol covers six greenhouse gases: CO_2, CH4, N_2O, HFCs, PFCs and SF6. Often GHG emissions are expressed as CO_2. Although, technically speaking, GHG emissions are expressed in CO_2eq, CO_2 can be used as a rough proxy. CO_2 emissions are easier to calculate and monitor than many others. In addition CO_2 is by far the most important source of GHG emissions. CO_2 contributes approximately 80% of total global GHG emissions, methane 14% and N_2O 8%, while the industrial gases account for 1% of total GHG emissions (see Figure 1.1).

While most of the CO_2 emissions stem from fossil fuel burning, a considerable share – roughly one-fifth of that – is emitted by land use, land-use change and forestry (so-called 'LULUCF'), i.e. deforestation, forest degradation or agricultural land degradation. If one included agriculture and deforestation fully in emissions budgets, Indonesia e.g. would rank among the top-five emitting countries of the world. Similarly, for Brazil, CO_2 emissions from land-use changes, i.e. deforestation, account for 40% of the total. In developed countries CO_2 from fossil fuel burning accounts for 80% on average.

Figure 1.1 Share of different anthropogenic GHGs in total world emissions in 2004 in terms of carbon dioxide equivalents (CO_2-eq)

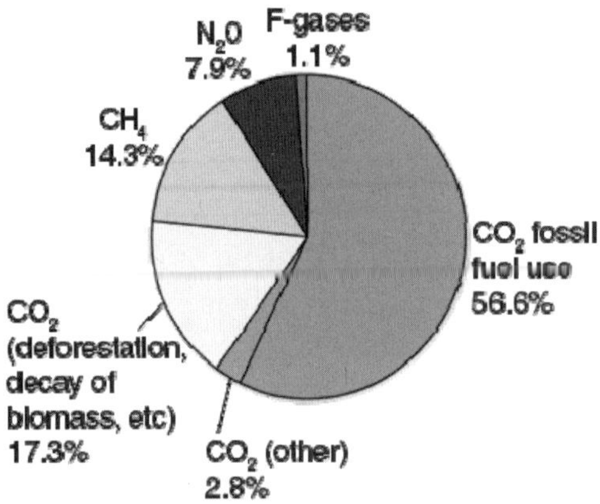

Source: Adapted from IPCC (2007, p. 5).

CH_4	Methane
N_2O	Nitrous oxide
CO_2 (deforestation, decay of biomass, etc.)	Carbon dioxide from land-use change and forestry etc.
CO_2 fossil fuel use	Carbon dioxide from fossil fuel use (includes emissions from feedstocks)
CO_2 (other)	Carbon dioxide from cement production and natural gas flaring
F-gases:	
SF_6	Sulphur Hexafluoride
PFCs	Perfluorcarbons
HFCs	Hydrofluorcarbons

1.1 The EU as the first mover

There is general agreement among economists (and other experts) that the most cost-effective way to lower CO_2 emissions is to have a 'carbon price', i.e. to have a system under which emitters pay for each tonne of CO_2 they emit. The EU has already put in place, and has been running since 2005, the most important carbon trading system in the world, the EU Emissions Trading System (EU ETS), which provides such a price. The EU ETS is a cap-and-trade scheme: it caps the total GHG emissions from the covered sectors and allocates allowances to emit that can be traded within the scheme. Thus the ETS establishes a benchmark carbon price which covers somewhat less than 50% of EU CO_2 emissions from power and industry sectors, including process emissions.

The Kyoto Protocol set up three flexible mechanisms: the Clean Development Mechanism (CDM), Joint Implementation (JI) and international emissions trading (AAUs trading), which at the time of the agreement had been seen as a nucleus of a global carbon market and price. The CDM is a mechanism that allows the creating of credits (Certified Emissions Reductions or CERs) in developing countries to be generated by investment in carbon-reduction projects to offset emissions in capped countries. The JI is a similar mechanism but applicable to developed countries with a cap. The reason for this mechanism has been that what were then Economies in Transition (EITs), such as Russia or Ukraine, have been allocated excess emissions rights, thereby removing any incentive to reduce economy-wide emissions. Finally, international emissions trading describes the possibility for governments to 'trade' or better 'exchange' the emissions rights that have been allocated to them under the Kyoto Protocol. While the CDM has been a moderate success in stimulating some low-carbon investment in developing countries, it hardly could fulfil the objective to create a global carbon price given its limited scale. To date, all CDM projects together are reckoned to have generated some 450 million tonnes of reductions of CO_2eq annually. This is small compared to global emissions of about 49 billion tonnes of CO_2eq.

International Emissions Trading has been even more limited. As a government-to-government mechanism, it could never create a global carbon market given the limited number of participants. While the current climate change negotiations are considering scaling up the volume of flexible mechanisms essentially by streamlining existing mechanisms, CDM and JI, and adding new ones aimed at expanding the scope of emissions sources, it is highly unlikely that these mechanisms will achieve the scope required to provide a viable carbon price signal.

The EU has decided that it will keep and expand in scope the trading scheme under the so-called 'post-2012 regime'. The EU ETS for the third phase (2013-20) will significantly contribute to the EU 20-20 targets: by the year 2020, total GHG emissions not only from ETS but also non-ETS sectors should be reduced by 20% compared to the 1990 level. This commitment translates into a 21% reduction from 2005 for the ETS sector. For more details, see chapter 2. EU targets outside the ETS sectors are not always underpinned by a general price signal, but are supposed to be reached by a wide variety of national subsidies (e.g. renewable energy feed-in tariffs or quotas) and direct regulations (e.g. on energy efficiency). However, some member countries (e.g. Sweden and possibly also France) have also introduced wider carbon taxes that cover all non-ETS consumption of fossil fuels, thus implying that practically all domestic sources of CO_2 are subject to a price signal.

1.2 ... to be reproduced in the US?

Until some time ago, it was considered as very likely that the US would soon take similar steps (probably with a cap-and-trade system) under the

Obama administration. Whether this will happen any time soon is difficult to predict now, however. The American Clean Energy and Security (ACES) Act (the Waxman-Markey bill), which the House passed in June 2009, envisages the establishment of an economy-wide GHG cap-and-trade scheme, including a number of cost-containment measures such as offsets. However, this act still has to pass the Senate and will likely be changed substantially before it is approved. However, there is wide agreement in the US that any US cap-and-trade system would be complemented by 'border measures' to level the playing field and limit carbon leakage.

1.3 But what about the rest of the world?

The EU and the US together account for about one-third of global emissions (and a similarly large share of global oil consumption), but they alone cannot deal with the problem since the emerging economies are quickly catching up in terms of CO_2 emissions as well (with China alone already today a larger source of CO_2 emissions than the EU or the US).

Since EU leaders seem determined to go ahead with some stringent post-Kyoto regime and the US is likely to follow suit, the key issue for European policy-makers (and industry) will thus be the participation of emerging economies in carbon pricing.

This problem has both a micro- and macroeconomic dimension.

The micro dimension concerns the impact of the EU (and the US) on specific, energy-intensive industries. The representatives of industry in Europe have so far taken the position that they support cap-and-trade systems, but that industry must be given free CO_2 allocations as long as major competitors, including emerging economies, do not face a similar carbon price. These concerns have been largely taken on board as over 160 manufacturing sectors, accounting for almost 80% of emissions, are likely to receive free allocations because they are considered at significant risk of leakage.

However, the free allocation of allowances represents merely redistribution within the EU. The shadow price of carbon will remain and would even do so under – hypothetical – *full* free allocation to most of manufacturing as long as the emissions allowances can be traded and fetch a positive price. In this case, even free allowances entail an opportunity cost. This implies that production and hence carbon leakage could still substantially undermine the effectiveness of the ETS.

As shown below, there are strong reasons to believe that unilateral measures by the EU or the US will be of limited usefulness (and may possibly even be counterproductive, or very costly) as long as major emerging economies, such as China, do not put in place equivalent measures. Concerns about carbon leakage will not go away until the latter makes a voluntary move.[2] This is why in both in the EU and the US there is a discussion on 'border measures', i.e. carbon tariffs against 'non-participating' countries.

The macro issue is that the relative weights among the major emitters are changing quickly. With the EU and US (and other OECD countries) likely to agree on reduction (targets) of around 20% or more by 2020 while emissions in emerging markets continue to grow exponentially, it is clear that attention has to shift to the major future emitters in the developing world. Given the dominance of China among the emerging economies, both in terms of the absolute level of its emissions and its growth rate, the key issue is thus essentially how to prevent 'China leakage'.

It is widely feared that 'border measures', as they are called, would lead to a major disruption of the global trading system because they could lead to a trade war. This would, of course, be most unwelcome especially in the context of the current financial crisis. However, this need not be the case. First of all, a border measure would probably be compatible with WTO rules, as shown in this study. Thus the EU's partners would not have any legal basis for retaliation. Moreover, the negotiations that are taking place following the UN conference in Copenhagen in December 2009 offer opportunities to negotiate another solution to 'leakage', namely the introduction of equivalent measures in China and other emerging

---

[2] A high rate of leakage could actually constitute another argument for unilateral measures. Sinn (2008), for example, argues that the only impact of a unilateral carbon price in the EU (and the US) would be to lower the price for fossil fuels without any impact on the global consumption of fossil fuels and thus without any impact on global CO_2 emissions. In other words, without border measures there might be 100% leakage. However, even if one accepts this argument, the EU (and the US) might still benefit from the lower price for imported hydrocarbons (oil and gas). A carbon tax at the border would at least serve to lower oil prices even further. It is doubtful, however, that the argument advanced by Sinn applies in reality. In fact, a carbon tax might actually lead to higher prices for hydrocarbons, which generate much less CO_2 per energy unit than coal.

economies. The pressure to do so will increase when the US unveils its own climate change policy, which is widely expected to contain border measures as well.

It is clear that it would be much better if China and other major emerging markets could be persuaded to take their own domestic measures to establish a price for carbon, thus avoiding 'border measures' altogether.[3]

In the context of the Copenhagen Accord, the Chinese government confirmed its unilateral pledge in a letter to the UNFCCC Secretariat that "China will endeavour to lower its carbon dioxide emissions per unit of GDP by 40-45% by 2020 compared to the 2005 level".[4] At first sight, this appears to constitute a significant commitment. It is not clear, however, whether this implies a major departure from the baseline because the emissions intensity of the Chinese economy should be falling in any event, as services become relatively more important. Schmidt & Marschinski (2009) even suggest that emissions under the target may in fact be higher than the business-as-usual projections[5] when existing and recently announced measures are taken into account, while the level of absolute emissions in 2020 also depends on the GDP growth assumptions (see also Chapter 5). It is thus difficult to say whether this target implies a meaningful price for carbon. Moreover, it should be pointed out that this is a unilateral target and the extent of any international monitoring of China's compliance is unclear. 'Unilateral' implies that it might be changed and the potential lack of international monitoring means that it will be difficult to verify whether it will actually be reached.

---

[3] However, from the point of view of the EU, it might be preferable to have border measures because the EU could then keep the tariff revenues, which could be substantial. With a carbon tariff of close to 8% (see chapter 7) and EU imports from China around €180 billion, the tariff revenue could amount to €14.4 billion per annum on Chinese imports alone.

[4] The letter is available at the UNFCCC website (at http://unfccc.int) under the heading "Appendix II – Nationally appropriate mitigation actions of developing country Parties".

[5] Projections by IEA (2009b) and EIA (2009) are used, compared to a -42.5% emissions intensity target (Schmidt & Marschinski, 2009, p. 11).

All in all, it thus appears that it will be difficult to determine whether there is at least an implicit pricing of carbon in China (and other major emitters).

1.4 Tariffs are always bad?

In the academic literature and in the WTO community, there is a strong general aversion to the use of any border measures, which are, *a priori*, seen as leading to distortions. The academic community has so far argued mostly that border measures are either not really necessary or are largely intended to protect special interest groups, which should be avoided in general. However, the terms of the debate have so far focused on the wrong issues. The basic reason why border measures are desirable is not so much because they 'level the playing field' (implicitly to protect domestic energy-intensive industries). Rather, they are desirable from a global welfare point of view because they will establish a 'shadow' carbon price even in countries that do not take any domestic measures to tax CO_2 emissions.

If the EU (or even more so, if both the EU and the US) were to introduce import tariffs based on CO_2 content, the production of CO_2-intensive goods in China would diminish (and that of other products would increase). Border measures in the EU and the US are thus useful in lowering the production of CO_2-intensive goods relative to those goods with a lower CO_2 footprint in the rest of the world. This implies that the imposition of a border carbon tax by the EU would increase global welfare. This argument has so far not been recognised in the literature.

1.5 The physics and politics of different fossil fuels: Hydrocarbons versus pure carbon (coal)

A key factor affecting the politics of the global discussions on what to do about climate change is the fact that the impact of a given price of CO_2 on the demand for different fossil fuels depends on how much each fuel already costs in terms of a unit of CO_2 emitted when it is burned. There is a huge difference in this respect between hydrocarbons (oil and gas) and coal.

The best way to think about the impact of a carbon tax on different fuels is to first calculate how much each fuel costs in terms of tonnes of CO_2 released (when burned). Table 1.1 below shows the relevant parameters at current prices. Column 1 shows the unit price (at the end of 2009). The prices of coal and gas are of course location-specific. We have used here average prices for several locations.

Column 2 then shows the price per 'tonne of oil equivalent' (toe), which can also be expressed as the price per unit of heat released (gigajoules). Column 4 then shows the price per unit of emissions, taking into account the fact that different fuels release different amounts of CO_2. Hydrocarbons derive some of their energy from the oxidation of hydrogen. Thus they release much less CO_2 per unit of energy created than coal. On the decisive metric of cost per unit of emissions, coal costs only around $34 (per tonne of CO_2 released), compared to over $150 for oil and gas.

This implies that a carbon tax of around $30 per tonne of CO_2 would almost double the price of coal, but would increase that of hydrocarbons, such as oil and gas, by as little as 20% or even less.

Table 1.1 Calculations for price per tonne of CO_2 for major fossil fuels ($)

	(1) Unit price	**(2) Price per toe**[a]	**(3) Price per GJ**[b]	**(4) Price per tonne CO_2**	**(5) Increase in cost with $30 CO_2 charge**
Oil (per barrel)	75	550	13.1	179	17%
Coal (per tonne)	90	135	3.2	34	88%
Gas (per BTU)	8.9	48.3	9.4	166	18%

[a] toe = tonne of oil equivalent

[b] 1 toe equals 42 gigajoules.

Source: Own calculations based on BP statistics.

At present prices, one would thus expect little impact on the demand for oil and gas from the imposition of widespread carbon pricing. However, the use of coal would become much more expensive. This implies that countries that rely heavily on coal would be relatively more affected by widespread carbon pricing than countries relying more on oil and gas. Unfortunately, the world's two largest emerging markets, India and China, rely on coal for most of their electricity generation and for their rapidly expanding steel production. To make matters worse, the coal burnt is mainly domestic. This might be the key reason why they have refused any explicit carbon pricing.

The absence of an explicit price for carbon throughout the economy, or at least in the sectors producing exports, is a key issue as shown in Chapter 3. Before going into this issue, Chapter 2 discusses briefly the framework for carbon pricing in the EU.

2. EU CLIMATE CHANGE STRATEGIES FOR THE POST-2012 PERIOD

This chapter establishes the context for the discussion about a potential EU carbon border tax. The focus is on the structure of the EU's own strategy to limit emissions through the EU ETS, a 'cap-and-trade' system and other measures.

2.1 The strategic approach

As early as 1996, the EU adopted a long-term target of limiting the temperature increase to a maximum of 2°C above pre-industrial levels. The EU target was reiterated over the years, most recently by the European Council of 18-19 June 2009 (European Council, 2009, p. 11) and laid the basis for domestic policies and measures aimed at mitigation of, and adaptation to climate change.[6]

The EU's stance must be understood in the context of the multilateral negotiations where the EU has traditionally played an important role. This is also the case regarding environmental issues. The EU as well as its member states have been actively promoting Multilateral Environmental Agreements (MEAs). In the case of climate change, the EU has been catapulted into a leadership role after President George W. Bush pulled the US out of the Kyoto Protocol. While few would have bet at that time that the Kyoto Protocol would survive, active EU diplomacy ensured that Japan, Canada and Russia ratified the Protocol, which entered into force in 2005. The EU adopted numerous legal texts to fulfil its commitments, including policies to support renewable energy and to improve energy

[6] The following sections in this chapter draw from Egenhofer et al. (forthcoming).

efficiency in buildings and transport. However, the centrepiece of EU climate change policy has been the EU Emissions Trading Scheme, which became operational in 2005 (see sections 2.5 and 2.6). While these and other policies have focused on the implementation of the Kyoto Protocol commitments, in parallel the EU has been developing a new strategy to meet mid- and longer-term climate change objectives, complemented by legislation (see sections 2.2 to 2.4).

An integral part of this strategy has been energy supply security and industrial policy considerations. The EU is facing changing conditions in energy supply: domestic energy resources are dwindling at the same time that government intervention in the energy industry is on the rise in precisely those countries that could potentially fill the gap. While many supplier countries such as those from OPEC or Russia seem unable to increase production due to a lack of investments, the fact that supplies are tightly controlled by governments in exporting countries raises the fear of 'excessive' leverage of supplier countries such as Russia. Many reserves will take years to develop due to problems of access, investments and physical conditions. A prolonged tight market might increase political tensions and possibly provoke some sort of 'resource nationalism'. In such a scenario, the EU and its member states have been examining domestic and external policy options to move to a more sustainable and secure energy supply. This includes, amongst others, investment in renewable energy sources, pushing carbon capture and storage (CCS) technology for fossil and other fuels and investment in nuclear energy in member states that wish to do so. To drive down costs for these technologies, there is a need for large-scale deployment. The International Energy Agency (IEA, 2008, p. 373 and p. 554) makes the case, for example, that renewables (except wind) experience significant capital cost reductions for each doubling of capacity, such as 15-20% for photovoltaics (PV) and 20% for solar water heaters. Pro-active support policies for low-carbon technologies are seen as a possible tool to gain leadership in low-carbon technologies. In addition, renewable electricity can reduce long-term electricity prices and their volatility; the substitution of fossils combined with renewables may reduce pricing power by Russia (notably on gas); and the introduction of the EU ETS has led to the retention of some of the economic rent of producer countries, including Russia. To offset the higher prices both for industry and domestic consumers, energy efficiency is a central piece of the strategy, certainly for the transition period until new technologies and new fuels become available on a large scale.

2.2 The climate and energy package

EU climate change policy is based on the EU's long-term target to limit global temperature increases to a maximum of 2°C above pre-industrial levels. The proposed range of a 20 to 30% cut has been derived from the recommendation of the 4th Assessment Report by the Intergovernmental Panel on Climate Change (IPCC, 2007). The report suggests that keeping the 2°C limit within reach would require atmospheric GHG concentrations to be stabilised at 450 parts per million (ppm), corresponding to emissions reductions from developed countries in the range of 25-40% by 2020 and 80-95% by 2050 (see section 5.3). The EU has suggested that at the same time developing countries as a group will need to limit the growth of their emissions to 15-30% below business as usual (European Commission, 2009a, p. 5). Research has shown that stabilisation at the required low concentration is technically feasible (see e.g. Neufeldt et al., 2009).

In order to achieve the medium-term GHG emissions reductions required of developed countries, the Council of the European Union formally adopted an integrated climate and energy package on 6 April 2009.[7] The package intends to operationalise the overall binding targets to reduce GHG emissions and to increase the share of renewable energy sources in the EU's energy mix, which were adopted by the European Heads of State and Government at their 8-9 March 2007 spring summit (European Council, 2007, pp. 12-21), generally referred to as '20 20 by 2020':

1) An absolute emissions reduction objective of 30% by 2020 compared to 1990, conditional on a global agreement with "comparable" commitments from other developed countries as well as adequate contribution by "economically more advanced developing countries", and a "firm independent commitment" to achieve at least a 20% reduction (European Council, 2007, p, 12);
2) a binding target to reach a 20% share of renewable energy sources in primary energy consumption by 2020;

[7] For a press statement on the Council's adoption of the 'climate-energy legislative package' as well as links to all of its elements, see http://www.consilium.europa.eu/uedocs/cms_data/docs/pressdata/en/misc/107136.pdf.

3) a binding minimum target of increasing the share of renewables in each member state's transport energy consumption to 10% by 2020;[8]

4) a 20% reduction of primary energy consumption by 2020 compared to projections (non-binding); and

5) a call for introducing a mechanism encouraging investment to enable the construction by 2015 of up to 12 large-scale power plants for CCS.[9]

2.3 Implementation

With the aim to implement these general targets, the climate and energy package, adopted on 6 April 2009, contains six key elements:

1) a revised EU ETS starting in 2013, which will bring about a 21% emissions reduction compared to 2005 in sectors covered by the EU ETS (European Parliament/Council of the European Union, 2009a, pp. 63-64);

2) an 'effort-sharing' Decision that sets legally binding GHG emissions reduction targets in respective EU member states (ranging from -20% to +20%, see Table 2.1) for all sectors not covered by the EU ETS (European Parliament/Council of the European Union, 2009b) – such as buildings, transport, agriculture and waste which currently cover about 55-60% of EU emissions – amounting to an overall reduction of 10% below 2005 levels by 2020;

3) a Directive for the promotion of renewable energy sources, introducing differentiated binding national targets for the share of renewable energy sources in final energy consumption by 2020 (see Table 2.1), amounting to a 20% share for the EU as a whole (European Parliament/Council of the European Union, 2009c). The Directive includes a binding minimum target of a 10% share of renewable energy, including biofuels in all forms of transport by 2020;

4) a Regulation to reduce average CO_2 emissions of new passenger cars by 2015 to 120g/km and to 95g/km of the new car fleet by 2020;

---

[8] This target initially focused solely on biofuels but was later widened to include all forms of energy from renewable sources (see section 2.3).

[9] This formulation was first introduced in the June 2008 European Council conclusions (European Council, 2008, p. 12).

5) new environmental quality standards for fuels and biofuels (aimed at reducing GHG emissions from fuels by 6% over their whole life-cycle) by 2020; and

6) a regulatory framework for carbon capture and storage (CCS). In addition, the EU ETS Directive makes available a specific number of allowances from the new entrants' reserve for co-financing of up to 12 CCS demonstration plants until the end of 2015 (European Parliament/Council of the European Union, 2009a, p. 74).

Prior to that, the EU had already published the so-called 'Strategic Energy Technology' (SET) Plan (European Commission, 2007) to strengthen research, development and demonstration of new technologies including those relevant for addressing climate change, and that is now being implemented.[10] Finally, a review of the level and nature of allowed subsidies (or 'state aid') is ongoing.

However, at the heart of the agreement are the '20 20 by 2020' targets. A single EU-wide cap is placed on GHG emissions from the ETS sector. Targets on GHG emissions from non-ETS sectors and on a renewable share are set for and differentiated by member states (see Table 2.1).

EU agreement on hard emissions caps and binding renewable obligations for member states has been based on a complex burden-sharing system. Hard targets for the EU ETS and the non-ETS sectors as well as for renewables have been set on the basis of an 'efficiency approach', i.e. reflecting a least-cost approach for the EU as a whole, but with some adjustment to ensure that costs for member states remain roughly similar in per-capita terms. The methodologies for the three targets are as follows:

- *GHG reduction ('effort-sharing') targets.* Countries with a low GDP per capita are allowed to emit more than they did in 2005 in non-EU ETS sectors, reflecting projected higher emissions due to higher economic growth. According to European Commission modelling, this increases overall EU compliance costs for the 20% GHG reduction target by 0.03% of total EU GDP.

---

[10] In October 2009, the European Commission (2009e) released a Communication on Investing in the Development of Low Carbon Technologies (SET-Plan) and accompanying documents, including an impact assessment, a technology roadmap and a document on R&D investment in the SET priority technologies.

Table 2.1 National overall targets for the share of energy from renewable sources in gross final consumption of energy in 2020 and GHG emissions limits in non-ETS sectors for the period 2013-20

Member state	Share of energy from renewable sources in gross final consumption of energy, 2005	Targeted share of energy from renewable sources in gross final consumption of energy, 2020	Member state GHG emissions limits in 2020 compared to 2005 GHG emissions levels (from sources not covered by ETS)
Austria	23.3%	34%	-16%
Belgium	2.2%	13%	-15%
Bulgaria	9.4%	16%	20%
Czech Rep	6.1%	13%	9%
Cyprus	2.9%	13%	-5%
Denmark	17%	30%	-20%
Estonia	18.0%	25%	11%
Finland	28.5%	38%	-16%
France	10.3%	23%	-14%
Germany	5.8%	18%	-14%
Greece	6.9%	18%	-4%
Hungary	4.3%	13%	10%
Ireland	3.1%	16%	-20%
Italy	5.2%	17%	-13%
Latvia	32.6%	40%	17%
Lithuania	15.0%	23%	15%
Luxembourg	0.9%	11%	-20%
Malta	0%	10%	5%
Netherlands	2.4%	14%	-16%
Poland	7.2%	15%	14%
Portugal	20.5%	31%	1%
Romania	17.8%	24%	19%
Slovak Rep	6.7%	14%	13%
Slovenia	16.0%	25%	4%
Spain	8.7%	20%	-10%
Sweden	39.8%	49%	-17%
UK	1.3%	15%	-16%

Sources: European Parliament, Council of the European Union, 2009b, p. 147; and European Parliament, Council of the European Union, 2009c, p. 46.

- *Renewables targets.* Half calculated on a flat-rate increase in the share of renewable energy and the other half weighted by GDP, modulated to take account of national starting points and efforts already made.
- *In the EU ETS sector.* Uniform cap across member states and allocation based on EU-wide allocation methodologies. Some 12% of the overall auctioning rights will be re-distributed to economically weaker member states in Central and Eastern Europe. Another 2% of the total auctioning rights will be distributed to eight countries that have already achieved significant reductions before 2005.

In addition to the equity questions, this elaborated impact assessment has also been required to examine larger macroeconomic and security of energy supply issues.

2.4 Costs of the integrated package

The European Commission (2008a, pp. 22-25; 2008b, pp. 159-163) has estimated the total direct costs of implementing the two binding targets for GHG emissions and estimated renewables at 0.6% of the GDP in the year 2020, or some €90 billion. However, through the access to offsets under the Clean Development Mechanism (CDM) and Joint Implementation (JI) both in the ETS and non-ETS sectors, compliance costs were expected to fall to 0.45% of GDP in 2020 – or roughly €70 billion (European Commission, 2008b, p. 161). Rising oil prices would also contribute to lower costs. Annual GDP growth is estimated to decrease by approximately 0.04-0.06% between 2013 and 2020, which would lead in 2020 to a GDP reduction of 0.5% compared to a 'business-as-usual' scenario. These calculations do not take into account possible macroeconomic benefits (in the estimated magnitude of +0.15% of GDP) from the re-injection of auctioning revenues back into the economy.

In the non-ETS sectors (covering about 55-60% of EU GHG emissions), the package permits member states to meet up to two-thirds of their emissions reductions by offset credits generated from CDM and JI, i.e. CERs (certified emissions reductions) and ERUs (emissions reduction units), respectively, allowing eleven countries to use additional offset credits, and the remaining part by domestic abatement measures.

2.5 The EU ETS and the development of international carbon markets

Since the adoption of the original EU ETS Directive in 2003, a broad consensus has emerged in the EU to use carbon pricing in the form of emissions trading, i.e. a cap-and-trade scheme, as the foundation of its climate policy. If properly designed, a cap-and-trade programme creates incentives for companies to reduce emissions in the most cost-effective way, rewards carbon-efficiency and creates incentives for new and innovative approaches to reduce emissions. The incentive for efficient abatement will arise from the 'opportunity costs' of using allowances. Passing through the costs of GHG emissions allowances to consumers will create incentives to reduce the demand for GHG-intensive goods. At the same time, this will increase producers' cash flow to invest in abatement technologies.

If all competitors were subject to similar carbon constraints in well functioning markets, the EU ETS would be the most suitable tool to achieve EU and UN-based targets at the lowest possible costs. However, the original design of the system had several flaws, partly but not only because of the absence of a global carbon price or at least a global climate change agreement.

The pilot phase from 2005-07 suffered from a number of teething problems such as significant delays of registries and National Allocation Plans (NAPs),[11] inconsistencies in the definitions of installations, as well as issues related to monitoring, reporting, verification and data collection. However, the most severe deficiencies of the first phase of the EU ETS included over-allocation, intensifying the effects of free allocation, distorting allocation between member states and generating windfall profits for the power sector.[12]

- The existing rules on free allocation of allowances has led to power companies charging their consumers as if they were paying a carbon price regardless of electricity sources, resulting in billions of windfall

---

[11] Some NAPs were delayed as much as 1.5 years.

[12] See Matthes et al. (2005), Swedish Energy Agency (2006), Ellerman et al. (2007), Egenhofer (2007), Ellerman & Joskow (2008) and Ellerman et al. (2010).

profits, estimated to amount to as much as €13 billion annually (Martinez & Neuhoff, 2005, p. 67).

- One reason for over-allocation was the absence of a hard constraint which led to inflated projections (AEA Technology Environment/ Ecofys UK, 2006). The combination of modest cuts and inflated projections resulted in over-allocation of as much as 97 Mt of CO_2 out of a total of about 2.2 billion annual EU allowances, i.e. almost 5% of total annual allowances (Kettner et al., 2007).
- Another reason for over-allocation was an excessive degree of decentralisation in the implementation of the EU ETS. This high degree of discretion for member states increased complexity, administrative burdens and transaction costs while decreasing transparency. It also enabled industry to put pressure on governments to hand out as many allowances as other governments do.[13]

Over-allocation has been addressed for the second phase (2008-12) when the European Commission could impose a formula[14] to assess member states' allocation plans. As a result, the European Commission could shave off 10% of member states' proposed allocations. While the expected price for allowances had been around €20-25, the economic downturn since 2008 has made prices tumble. The use of the formula has been challenged before the European Court of Justice successfully by Poland and Estonia, therefore necessitating a new NAP. It is uncertain at this moment what the consequences will be. There may not be any substantial ones, depending on the positions that member states in question take.

2.6 The EU ETS in the third phase (2013-20)

Experiences from the initial phases and design flaws have greatly helped the European Commission to propose radical changes to the EU ETS (see Egenhofer, 2009).

---

[13] See Zetterberg et al. (2004) and Matthes et al. (2005).

[14] Verified 2005 ETS emissions x GDP growth rates for 2005-10 based on the PRIMES model x carbon intensity improvements rate for 2005-10 + adjustment for new entrants and other changes, for example in ETS coverage.

The principal element of the new ETS is a single EU-wide cap which will decrease annually in a linear way, starting in 2013, to reach 1,720 million tonnes of CO_2 in 2020. This corresponds to an overall cap being 21% lower than the verified emissions for 2005. This linear reduction continues beyond 2020 as there is no sunset clause.

In addition, there are EU-wide harmonised allocation rules, full auctioning to sectors that can pass through their costs (e.g. the power sector) and partially free allocation to industry based on EU-wide harmonised benchmarks. Overall, this could by and large translate into 50% auctioning, which could equal about €27 billion per annum at a price of €30 per tonne of CO_2 (see Behrens et al., 2008).

Starting from 2013, power companies will have to buy all their emissions allowances at an auction. The auctioning rate in 2013 for existing power generators in some (mainly Eastern European) countries will be at least 30% and will be progressively raised to 100% thereafter. This means, e.g. that existing coal-fired power plants in Poland still get their allowances for free, but that new power plants need to buy them.

For the industrial sectors under the ETS, the EU agreed that the auctioning rate will be set at 20% in 2013, increasing to 70% in 2020, with a view to reaching 100% in 2027. Industries exposed to significant non-EU competition, however, will receive 100% of allowances free of charge up to 2020, based on Community-wide benchmarks. The latest indications suggest that the majority of manufacturing industry falls into this category (see Box 4.4 in Chapter 4).

Furthermore, 12% of the overall auctioning rights will be re-distributed to lower per capita member states (10%) and those that have undertaken early action (2%). The system will be extended to the chemicals and aluminium sectors and to other GHGs (e.g. nitrous oxide from fertilisers and perfluorocarbons from aluminium). The EU ETS will also include aviation from 2012 onwards. This has raised major disputes with the US administration including those on trade matters.

The left-over CDM/JI credits from 2008-12 can be used until 2020. This amounts to approximately 2 billion tonnes.

The EU ETS Directive allows for linking the EU ETS with other emissions trading schemes by different types of linking arrangements, e.g. via an international treaty, an international agreement as foreseen under EU law and through a reciprocal commitment applied through domestic systems. In essence, over time the EU ETS could link non-EU emissions trading schemes, thereby developing the global carbon market.

There is little doubt about the ETS and its enforcement: EU commitments are legally binding under EU law and by and large enforceable by the European Commission. Participating companies will need to surrender allowances or face penalties and ultimately closure. The European Commission can take member states to court if they do not meet the absolute reduction targets on GHG emissions from non-ETS sectors. Irrespective of a UN-based compliance mechanism, the EU and its member states are bound by EU law to achieve their GHG reduction targets. While in theory, the EU decision could be reversed, in practice this is extremely unlikely given the complicated EU decision-making processes.

2.7 Concluding remarks

This chapter shows the complications brought about by the absence of a global emissions trading market. It highlights the compromises that have been struck to cope with issues such as carbon and production leakage, the latter being a politically delicate issue. An extension of the 'cap-and-trade' approach to the most important emitters would be of course desirable, but unfortunately none of the emerging economies, including China, have given any sign of willingness to commit to significant reductions in emissions, let alone, to adopt the 'cap-and-trade' approach. A global emissions trading system that would encompass these major sources of emissions thus seems a long way off. This raises the central question addressed in the remainder of this study: Would a carbon border tax or tariff be useful and justified under these circumstances?

3. Global Welfare Implications of Carbon Border Taxes

This chapter focuses on the economic mechanisms that allow one to compute the welfare consequences of the introduction of a tariff on the CO_2 content of imported goods in a country that already imposes domestic carbon tax. The main finding is that the introduction of a carbon import tariff increases *global* welfare (and not just the welfare of the importing country) if there is no (or insufficient) pricing of carbon abroad. A higher domestic price of carbon justifies a higher import tariff. Moreover, a higher relative intensity of carbon abroad increases the desirability of a high import tariff being imposed by the importing country because a border tax shifts production to the importing country, which in this case leads to lower environmental costs.

If both instruments, i.e. import tariffs and domestic carbon prices, are used to maximise global welfare, the optimal domestic price for carbon should be higher than the external effects (assuming that there is no carbon pricing in the rest of the world) and the optimal tariff rate would be somewhat lower than the domestic carbon price.

If the importing country has a fixed ceiling on emissions instead of a constant carbon price (as provided under the EU Emissions Trading System), an import tariff is always beneficial from a global point of view and its imposition lowers the domestic carbon price, but less than proportionally.

3.1 Introduction

The costs and benefits of 'border measures' have been extensively discussed in the rapidly growing literature on the economics of climate change mitigation policies, but most studies concentrate on competitiveness (of energy-intensive industries) and carbon leakage (see e.g. Trouser et al., 2008; Veenendaal & Manders, 2008; Mc Kibben & Wilcoxen, 2008; Frankel, 2009). Only a few studies examine the international trade impacts of a 'carbon border tax' (e.g. Hufbauer et al., 2009) and none looks at the welfare implications from a global point of view.

The purpose of this chapter is thus to provide a solid basis for any discussion of the economics of 'border measures' to combat climate change. Since climate change policy, even when implemented at the national level, is motivated by a concern for global (as opposed to national) welfare, it is important to adopt the same point of view when discussing so-called 'border measures'.

An important side issue in the discussion about 'border measures' is the distinction between plain import tariffs (on the carbon content of goods imported) and the combination of import tariffs plus export rebates.[15]

This chapter focuses on the case where there is no export rebate. The model used here has only one good (of which the importing country is a net importer), and hence it cannot be used directly to assess the impact of the combination of an import tariff plus an export subsidy. However, this should not be of major importance given the well-known general result from the theoretical literature: a generalised (ad valorem) export subsidy coupled with an import tariff is equivalent to a depreciation of the nominal exchange rate and thus has no impact in the long run when all nominal variables can adjust.

---

[15] The combination of import tariffs and export refunds constitutes what is usually referred to as 'border tax adjustment' or BTA. Of course, the concept of border tax adjustment is not new, but its application to environmental problems is. According to the final report of the decisive GATT Working Party (1970), a BTA is defined "as any fiscal measure which puts into effect, in whole or in part, the destination principle". The destination principle enables exported products to be reimbursed for some or all of the taxes charged in the exporting country and imported products to be charged with some or all of the taxes charged in the importing country (GATT Document L/3464). Furthermore, the Working Party concluded that only certain indirect taxes but not direct taxes (such as social security charges and payroll taxes) were eligible for tax adjustment. This conclusion was important for the EU as BTAs are widely used by the EU owing to the fact that member countries rely on indirect taxes (VAT).

By contrast, there is no consensus in the empirical literature whether a border tax adjustment is effective or not, i.e. whether it can correct for the distortionary impacts of (national) climate mitigation policies that result in a loss of competitiveness and carbon leakage.

A study by Veenendaal & Manders (2008) addresses directly the effectiveness of a carbon BTA on competitiveness and carbon leakage for the EU, assuming that the EU is the only country to follow this approach. In a general equilibrium analysis, they quantify the impact of a number of policy scenarios with a specific focus on the energy-intensive sectors covered by the EU ETS. They show that when there is no BTA (and no equivalent transfer mechanism in the form of a clean development mechanism or CDM), production and employment in these sectors are negatively affected by the imposition of a domestic tax on carbon (for example the ETS). However, the imposition of a BTA (but again no CDM) can mitigate the loss of competitiveness: the loss in production and employment is halved. The overall welfare effects of the BTA for Europe is ambiguous since refunds are found to be welfare decreasing for Europe but import levies are welfare increasing. Overall the authors conclude that the impact of the BTA is too modest to make its implementation worthwhile. McKibben & Wilcoxen (2008) reach a similar conclusion: the benefits from the BTA are too small to justify their administrative complexity. In contrast, Majocchi & Missaglia (2002) use a general equilibrium model and show that BTAs are more likely to produce a better environment and less unemployment for the EU-15 member countries. Other empirical studies on sectoral impact give support to BTAs: Demailly & Quirion (2005) show that the BTA can be effective in preventing carbon leakage in the cement industry, whereas Mathiesen & Maestad (2002) follow a similar exercise for the steel industry.

The remainder of this chapter is organised as follows: The next section provides a simple illustration of the welfare gain from a carbon border tax based on a standard approach which ignores differences across countries in terms of their carbon efficiency. Section 3.3 then uses the same approach to illustrate a more specific point, namely that the pricing of carbon in the EU without a carbon border tax could actually lead to higher global emissions. A 'stand-alone' ETS might thus hurt global welfare rather than improve it. Section 3.4 concludes.

3.2 A simple illustration of the welfare gain from the introduction of a carbon tariff

This is a simple partial equilibrium illustration using linear demand and supply curves for simplicity to show graphically the impact of a carbon tax (i.e. a tariff on the carbon content of imports) on global welfare. As usual, the world is divided into two actors: an importing country (or group of importing countries) and the rest of the world (RoW).

Figure 3.1 shows the global demand and supply curves of the good in question. Production of the good leads to emissions of CO_2 (at a certain, given rate per unit). Private producers do not take into account the cost of emissions: hence the private supply curve is below the supply curve, which takes into account the external impact of emissions. However, unless there is further government intervention, the international price of the good in question is determined by the intersection of private supply and demand, point O on the graph.

Figure 3.1 Equilibrium without tariff

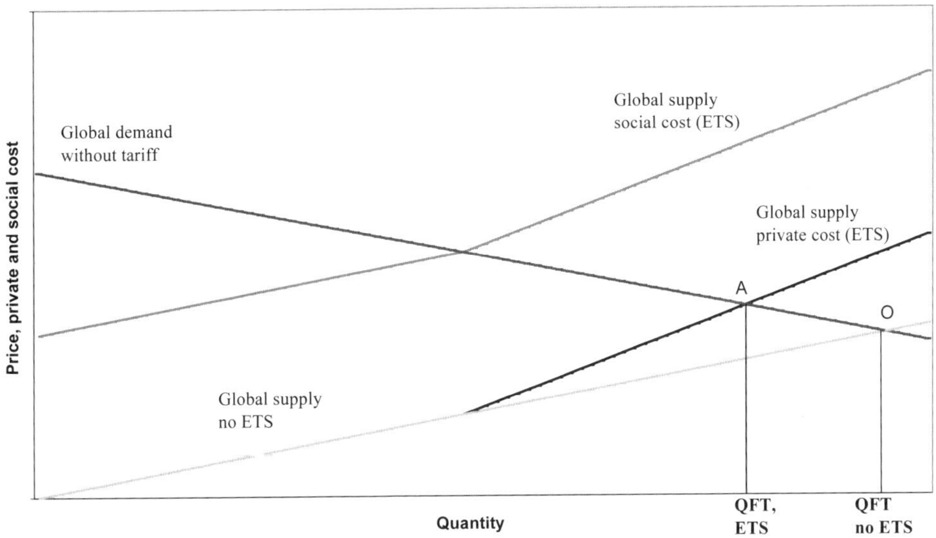

For simplicity, we consider here the simplest case, which is when the home country introduces a 'cap-and-trade' system, such as the ETS in Europe. In this case the global supply curve is 'kinked' at the quantity at which the ETS limits the amount that can be produced in the home country (given the limit on emissions and the per unit emissions factor). After the introduction of the ETS, the world price of the good in question is determined then at point A by the intersection between the ('kinked', private) supply curve subject and the global demand curve. It is apparent that the price is higher than before.

The issue at hand is what happens when the home country introduces an import tariff in addition to the ETS. The import tariff obviously reduces domestic demand and hence also global demand since it does not affect demand in the rest of the world. The import tariff is here assumed to be specific, not ad valorem, because it is supposed to correct the externality that arises in production. Figure 3.2 shows the resulting equilibrium: with the introduction of the tariff, the demand curve shifts down and the resulting new equilibrium price is lower than before (equilibrium shifts from point A to point E).

Figure 3.2 Equilibrium with tariff

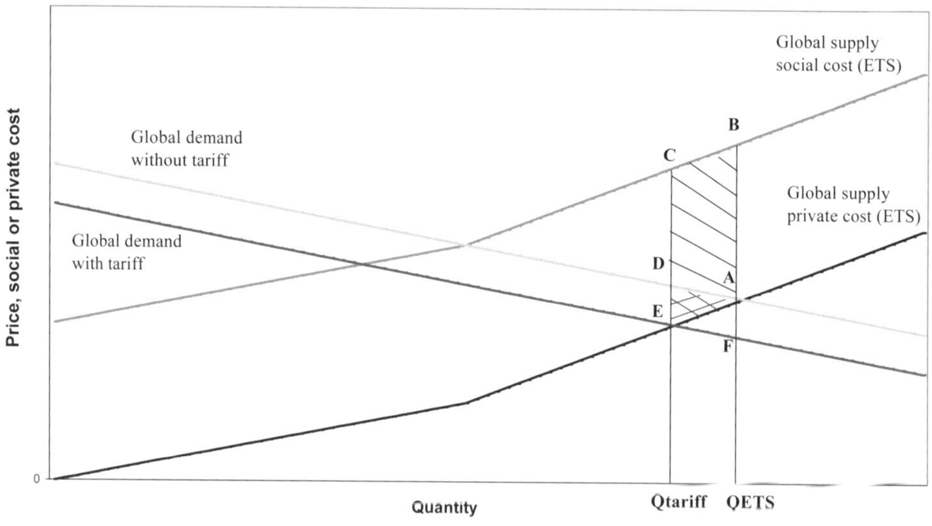

The fall in the international price implies that foreign producers will produce less. Given that domestic production is limited by the ETS, global production must fall as well. The sum of domestic plus foreign consumption must thus also fall. But this is achieved by a rise in foreign consumption (since the price falls abroad) and a fall in domestic consumption of an even larger magnitude. This shift in consumption is due to the fact that domestic and foreign consumers face a different price if there is an import tariff.

What are the welfare implications of the tariff? The standard welfare loss caused by a tariff is the usual triangle (consumer plus producer loss) enclosed by points ADE. As is well known, this welfare loss is of second order for any 'small' tariff.

In this case, however, there is also a gain due to the global externality in production. It is enclosed by the points ABCE (a 'rectangle' or parallelogram). The net welfare gain from imposing a tariff is given by the trapezoid enclosed by the points ABCD. It follows that a small carbon tariff must always improve global welfare.

The intuition behind this result is clear. As long as the tariff is small, the reallocation of consumption from consumers at home to consumers abroad causes only a loss of second-order importance. But the gain to global welfare from lower foreign production is of first-order importance.

3.3 In the absence of a carbon tariff, a cap-and-trade system can be counterproductive

Another way to illustrate the importance of a carbon import tax in rectifying the inefficiency created by the absence of carbon pricing abroad is to ask a simple question: Will the unilateral imposition of a 'stand alone' ETS, i.e. an ETS without border measures, always help the environment? The answer seems to be no. Some simple calculations show that it might not be unlikely in reality that an ETS without border measures (i.e. without a carbon import tariff) leads to higher global emissions. The intuition behind this result is straightforward.

It is clear that a binding cap on domestic emissions will restrict the domestic supply of all energy-intensive goods. This implies that the global price of these goods must increase, which in turn means that production abroad will increase, which of course will lead to higher emissions abroad. This phenomenon is called 'carbon leakage' in the parlance of the climate change community.

Most existing analysis of carbon leakage focuses on a small subset of energy-intensive sectors (steel, cement, etc.) whose products are often traded intensively. The EU has actually defined sectors exposed to a significant risk of carbon leakage mainly in terms of their openness to trade and found that about 60% of all ETS sectors (accounting for about 75% of emissions) are 'at risk'.[16] This implies that the potential for carbon leakage should be widespread. This impression is confirmed by a recent study based on a large general equilibrium model which concludes that about 40% of any reduction in the production of energy-intensive goods in the EU would be offset by higher production abroad.[17]

However, this focus in much of the literature on energy-intensive industries is misguided because it focuses on the wrong issue (the competitiveness of particular sectors) and neglects the fact that the output of these industries (especially energy and steel) is used throughout the economy. Most products that are traded intensively thus incorporate substantial amounts of emissions via the energy and the energy-intensive

---

[16] Article 10a of the revised Directive states that a sector or sub-sector is "deemed to be exposed to a significant risk of carbon leakage (CL) if:

- the extent to which the sum of direct and indirect additional costs induced by the implementation of this directive would lead to a substantial increase of production cost, calculated as a proportion of the Gross Value Added, of at least 5%; and
- the non-EU trade intensity defined as the ratio between total of value of exports to non EU + value of imports from non-EU and the total market size for the Community (annual turnover plus total imports) is above 10%."

A sector or sub-sector is also deemed to be exposed to a significant risk of carbon leakage:

- if the sum of direct and indirect additional costs induced by the implementation of this directive would lead to a particularly high increase of production cost, calculated as a proportion of the Gross Value Added, of at least 30%; or
- if the non-EU trade intensity defined as the ratio between total of value of exports to non EU + value of imports from non-EU and the total market size for the Community (annual turnover plus total imports) is above 30%.

The Commission's website (http://ec.europa.eu/environment/climat/emission/carbon_en.htm) reports that 151 of 258 NACE-4 sectors (≈ 60% of all sectors) face Significant Risk of CL (SRCL). Sectors deemed exposed to SRCL account for ≈ 75% of all GHG emissions by industries covered by the ETS.

[17] See Veenendaal & Manders (2008). Their study arrives, however, at much lower estimates for overall carbon leakage for reasons that are not clear.

inputs used in their production. Given that it is usually assumed that the supply of exports from China and other emerging market economies is rather price-elastic, even small changes in relative prices could have a considerable impact on trade flows.

It is well known that carbon leakage undermines the effectiveness of any national 'cap-and-trade' system, like the ETS, in reducing global emissions. However, it is not widely realised that under certain conditions carbon leakage could actually lead to the paradoxical situation in which the imposition of an ETS-type cap-and-trade system would lead to an increase in global emissions and thus a loss of (global) welfare.

The mechanism through which this can happen becomes clear once one distinguishes between 'production' and 'carbon' leakage. The former describes the displacement, at least partially, of domestic *production* to the rest of the world. The latter refers to the amount of *emissions* avoided when domestic production falls relative to the amount of carbon emitted in the rest of the world where production goes up. For example, if production leakage were to be only 50% (i.e. foreign production increases only by one half of the fall in domestic production), global emissions would still increase if the carbon intensity abroad is more than twice as high as at home. The general point is that the displacement of production, even if partial, can lead to an increase in overall emissions if the carbon intensity of production in the rest of the world is much higher than at home.

A key parameter in any judgment of the efficiency of the ETS (and the national carbon taxes in France and Sweden) is thus the difference in energy (and thus carbon) intensity between the EU and its major trading partners. How large is it? The best estimates of the emissions embodied in international trade are based on input-output matrices to take into account the way energy inputs are used throughout the economy. On this basis, Peters & Hertwich (2008) and Weber et al. (2008) suggest that (on average for all sectors) each $1,000 of exports from China contains about 2-3 tonnes of carbon, which is about four times more than the 0.5 tonnes of carbon embodied in $1,000 of exports from the EU or other OECD countries. The same sources also show that exports from other emerging markets have sometimes even higher carbon intensities than those of China.

Another way to provide a crude estimate of differences in carbon intensities can be gleaned from the carbon intensity of GDP. [18] This approach suggests that the carbon intensity of production is ten times higher in Russia than in the EU and six times higher in China. Of course, the higher carbon intensities of emerging markets are partially due to their different output mix. However, this does not change the basic fact that a USD unit increase in exports from China in general embodies four times as much CO_2 emissions as exports from the EU or the US. Moreover, higher exports in general lead to higher income and thus higher GDP and higher GDP growth in China is associated with a much higher growth rate of emissions.

Table 3.1 Carbon intensity of exports and GDP in selected key economies

	CO_2 intensity of exports	CO_2 intensity GDP 2005
EU27	0.47	0.43
US	0.72	0.53
China	2.46	2.43
India	2.67	1.78
Brazil	1.05	0.5
Russia	3.85	4.4

Source: Own calculations based on IFM data and Weber et al. (2008). The carbon intensity of exports is based on 2002 data. Both intensities are measured as tonnes of carbon per $1,000 of exports.

[18] The standard objection to the use of the data on carbon intensity per unit of GDP (or exports) is that it is naturally (also) a function of the economic specialisation of the country concerned. For example, Russia might be very carbon-intensive just because its specialisation is the extraction of hydrocarbons. Similarly the high carbon intensity of China is (also) a result of its specialisation in energy-intensive industrial goods. (Another reason might be the abundance of local coal and remaining energy subsidies.) However, a specialisation in energy-intensive industrial goods is no reason not to apply the simple rule that CO_2 emissions should be taxed wherever they occur and whatever the reason. CO_2 emissions should be taxed the same way whether they are produced in the manufacture of industrial goods or the transport of these goods or whether they arise from the heating of homes or leisure travel.

If one accepts as a benchmark that Chinese production is in general about 4 to 5 times as carbon intensive as that of the EU, it follows that the ETS might have led to an *increase* in global emissions if production leakage had been only somewhat above 20-25%.[19]

A proper welfare evaluation of the economic impact of a cap-and-trade system like the ETS (or of the more wide-ranging recent French proposal to tax all energy inputs, not only in the energy-intensive sectors) should also take into account the fact that the displacement of consumption and production leads to standard welfare effects. This is done in Figure 3.3 below, which shows the equilibrium conditions for a (possibly composite) good whose production creates emissions and thus an externality.

Figure 3.3 Effect of ETS with higher carbon intensity abroad

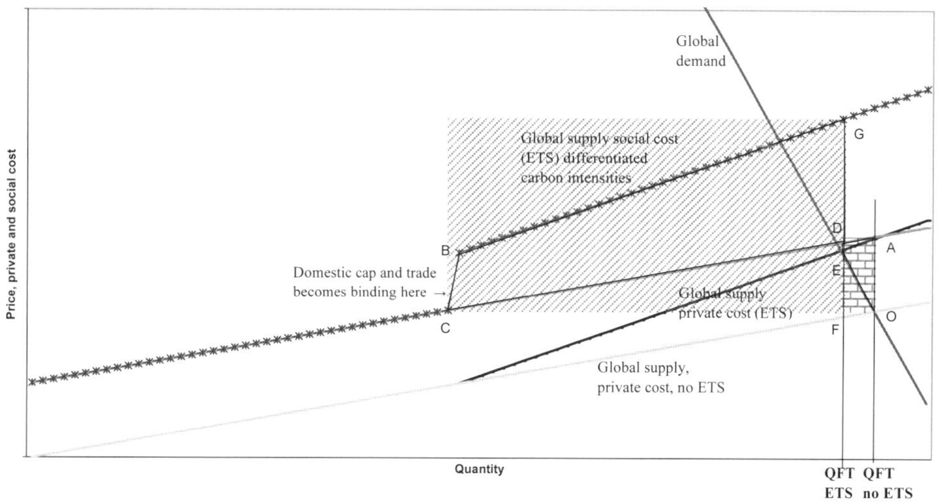

The introduction of a domestic carbon price (via a cap-and-trade system or by other means) has two effects that must be conceptually distinguished:

---

[19] It is of course impossible to determine the marginal carbon intensity for exports from emerging economies that are related to the imposition of the ETS in Europe. However, the burden of proof should be on those who argue that this marginal carbon intensity is much lower than the average measured by aggregate statistics.

1) It reduces global production, from $QFT_{no\ ETS}$ to QFT_{ETS}. This reduction in global production increases welfare because at the margin the social cost was higher than the (private=social) benefits from consumption. The net welfare gain is given by the area (covered by little rectangles) enclosed by the points OADE (loss of consumer surplus under the line OE but gain of social cost of production below the line DG).

2) The fact that the domestic price of carbon is higher than the price for carbon abroad leads to an increase in the social cost of production beyond the point at which the domestic cap or ceiling is reached. From this point onwards, the social cost is not only above the one for the unconstrained case (i.e. the case without a domestic cap on emissions) but also steeper because any additional production has to take place abroad. This increases the social cost for two reasons: first, the private cost of production is higher because the supply from domestic producers cannot increase. Secondly, the external effects from producing abroad are higher because the carbon intensity abroad is higher. This implies that the (global) social cost of producing the reduced quantity QFT_{ETS} is higher by the shaded trapezoid enclosed by the points BCGD.

As drawn, it is clear that an ETS-type cap-and-trade system can actually make the world worse off. Whether or not this is the case depends of course on the slopes of the demand and supply functions relative to the difference between domestic and foreign carbon intensities.

Gros (2009) shows in the context of a standard, fully specified model that a domestic price on carbon can be counterproductive in terms of global welfare under the following condition:

$$\frac{carbon\_intensit\_abroad}{carbon\_intensity\_at\_home} \equiv relative\_carbon\_intensity \rangle \frac{elasticity\_global\_demand}{elasticity\_of\_supply\_abroad}$$

As one would expect, carbon leakage is higher the lower the sum of the domestic and foreign elasticities of demand and the larger the foreign elasticity of supply, adjusted for the size of the foreign country. In other words: the introduction of a cap-and-trade system in a small country that has a much lower carbon intensity than the rest of the world has a high probability of being counterproductive. Moreover, the longer the time horizon, the higher should be the elasticity of supply. This implies that while carbon leakage might not be important in the short run, it could become much more relevant as time goes on.

This analysis would of course be relevant, *mutatis mutandis*, for the general carbon tax recently proposed in France, given that that country is small relative to the rest of the world. And given that the carbon intensity of the French economy is rather low, this measure could thus very well have a negative impact on global welfare. Adopting a domestic carbon tax at the EU level would not change the conclusion much since the EU also accounts for only a limited share of global GDP.

3.4 Model based analysis confirms these points

Gros (2009) shows that the simple intuition embodied in the graphs above remains valid in the context of a fully specified model of demand and supply of a good whose production involves an externality. He uses a simple partial equilibrium two-country model to show the impact on trade and global welfare of the combination of a domestic carbon tax with an import tariff. In this model the world is divided into two actors: an importing country (or group of importing countries) and the rest of the world (RoW), with potentially differing demand and supply functions as well as differing carbon intensities. Within this model one can analyse the impact of a 'border' tax (or import tariff) in the importing country on world production and welfare when production involves an externality (greenhouse gas emissions). The results of the model confirm the qualitative results illustrated graphically above.

Moreover, the model-based analysis also allows one to calculate the tariff rate that is best from the point of view of global welfare. This yields the following additional insights:

1) Not surprisingly, the optimal tariff is higher the higher the carbon intensity of production abroad. However, the optimal tariff remains positive even if the carbon intensity abroad is much lower than at home - provided of course that the foreign carbon price is insufficient to cover the externality abroad.
2) A higher slope of supply function abroad also leads to a higher optimal tariff.
3) A higher slope of the demand curve at home implies a lower optimal tariff.
4) Obviously a higher carbon tax abroad (ε^*) implies a lower optimal tariff.

5) The level of the domestic emissions cap does not directly influence the optimal tariff.

The analysis based on a formal model thus confirms and refines the general argument made here, namely that a tariff on the carbon content of imports is always welfare-improving in the presence of a domestic carbon price.

3.5 Concluding remarks

This chapter provides an evaluation of the economic impact of border measures in climate change policy from a global welfare point of view. Most of the literature has focused on two separate issues, namely carbon leakage and the loss of competitiveness of energy-intensive industries in countries that introduce limits on emissions. However, climate change policy is motivated in the first instance by a concern for global welfare. Hence one should not look at these issues from a national or regional point of view.

Moreover, addressing these issues from an economic welfare point of view means that one cannot look exclusively at the impact of any one measure on the amount of CO_2 emissions, but also at the cost of production and consumer welfare.

The results obtained with a simple standard approach are straightforward and intuitively clear. Overall, as one would expect, an import tariff improves global welfare because this transfers carbon pricing, at least partially, via trade flows, even to those parts of the world whose governments have so far refrained from imposing any domestic measures. The optimal level of the tariff rate (from a global welfare point of view) depends on many parameters, such as the elasticities of demand and supply, but the key parameter remains the external impact of emissions.

4. Compatibility of Carbon Border Measures with WTO Rules

This chapter investigates whether a carbon border tax be compatible with WTO rules? It turns out that the answer depends on a number of factors. We argue that the qualifying conditions can probably be met. Or, to put it differently, one can design an EU carbon border tax that is WTO-compatible.

WTO members are allowed to adopt trade-related measures aimed at protecting the environment. However, climate change is not part of the WTO's ongoing work programme and there are no WTO rules specific to climate change. Hence one has to look at WTO case law in conjunction with its general rules for an evaluation of the WTO compatibility of a carbon border tax. The answer depends on the details of its implementation.

We find in particular that the following points would be crucial to ensure WTO compatibility:

- The EU carbon border tax should cover all imports and be based on their carbon (CO_2) content. This has two important consequences: first, this is a tax on a 'process' not on a 'product', second, as this is a tax based on a 'process', it is origin-neutral, and hence does not violate Article I on most-favoured nation (MFN) treatment.
- As taxes based on 'process' violate Article III (national treatment of 'like' products), the EU has to invoke Article XX (general exceptions) (see Box 4.1). Under Article XX, trade-related measures are allowed if necessary to protect exhaustible resources or human, animal or plant life or health. The EU can argue that *a low-carbon atmosphere, necessary to avoid catastrophic climate change, should be viewed as an "exhaustible natural resource"* (as per Article XX (g)).

- The rate of tax should be set equal to the difference in the EU price of carbon (ε) and foreign price of carbon. Hence, the tax on a product is going to be equal to the carbon content times the difference in the EU carbon price and foreign carbon price.
- The carbon tax should be revenue-neutral: revenues collected from the carbon tax should be used to create a fund used to ease the transition of energy-intensive industries and to invest in new technologies.
- The methodology used to determine the carbon content of an imported product needs to be objective. An 'external' benchmark method such as the ISO 14067 might work.

Box 4.1 Article XX GATT/WTO: General exceptions

Subject to the requirement that such measures are not applied in a manner that would constitute a means of arbitrary or unjustifiable discrimination between countries where the same conditions prevail, or a disguised restriction on international trade, nothing in this Agreement shall be construed to prevent the adoption or enforcement by any contracting party of measures:

(b) necessary to protect human, animal or plant life or health;

(g) relating to the conservation of exhaustible natural resources if such measures are made effective in conjunction with restrictions on domestic production or consumption;

(h) undertaken in pursuance of obligations under any intergovernmental commodity agreement which conforms to criteria submitted to the CONTRACTING PARTIES and not disapproved by them or which is itself so submitted and not so disapproved.

The second bullet point is crucial as it provides the key justification. The arguments that increasing atmospheric concentrations of CO_2 damage the environment and that the absorption capacity of the earth of GHG emissions is limited are supported by science and numerous international declarations and agreements in the context of the activities of the UN on preventing global climate change (including the Kyoto Protocol). Action by the EU alone cannot be sufficient to prevent catastrophic climate change and efforts by major emerging market economies to curb CO_2 emissions are absolutely necessary.

Many of those who consider that border taxes are not compatible with WTO rules seem to overlook this point. See for example the following quote from Ahner (2009, p. 18):

> The crucial point is that the protection of domestic producers from foreign competition is not recognised as a legitimate policy objective under WTO law. For the time being, it does not seem that the EU will be able to credibly articulate that the measure is designed to achieve greenhouse gas reductions.

The key argument made here is that one can indeed show that its measures will achieve a reduction of GHG emissions.

As described in more detail in chapter 2, the key element of the EU's future climate strategy is the commitment to achieve at least a 20% reduction of greenhouse gases by 2020 compared to 1990 levels and in case of a conclusion of a comprehensive international climate change agreement, a 30% objective. The EU is also committed to honour its engagements under the Kyoto Protocol for which the EU has a specific tool that mainly affects industry (including power generation), namely the ETS, which has operated since 2005. Within this system some 10,000 energy-intensive plants can buy and sell CO_2 allowances. This system will be maintained in future to ensure compliance with the EU's targets by 2020.

The ETS increases costs for the sectors that are covered (namely power generation, iron and steel, glass, cement, pottery and bricks), which might erode the competitiveness of these sectors vis-à-vis countries that have no such stringent commitment to combat climate change, such as the US and emerging markets. Industry and business associations have naturally tended to emphasise this aspect. In the climate change literature the concern is not competitiveness, but 'carbon leakage', i.e. industries relocating to countries where there are no such climate change measures and hence a higher carbon intensity leading to the result of overall higher global carbon emissions.

At the political level, trade policy measures such as BTAs (border tax adjustments) have been supported, for example, by the French President Nicolas Sarkozy, German Environment Minister Sigmar Gabriel and Commission Vice-President Günter Verheugen (Quick, 2008). In its resolution on Trade and Climate Change, the European Parliament (2007) emphasises that the Commission should examine "WTO-compatible mechanisms and climate-friendly trade policies" to address such competitiveness and carbon leakage issues and "to consider that such trade measures should be taken only when alternative measures will be ineffective in achieving given environmental objective" and that they "should be no more trade-restrictive than necessary to achieve the objective". Consistently throughout the years, European Council Presidency Conclusions as well as conclusions from the Environmental Councils have listed border measures as a possible tool, if required.

Neither the idea of a border tax adjustment nor their use is new. In fact, border tax adjustments have been in use as early as the 18th century in the US. The difference today in comparison with earlier times is that the application of border tax adjustment is now rules-based thanks to the 1970 report of the Working Party on Border Tax Adjustment of GATT. The GATT ruling of 1970 on BTAs was initiated as a result of the adoption of Value Added Tax (VAT) in the EU in the early 1960s. As the academic literature at the time indicated that a VAT based on the 'destination' principle was trade-neutral, a border tax adjustment that imposed a tariff on EU imports and a rebate on EU's exports was to offer the EU no trade advantage (Lockwood & Whalley, 2008). The economic incidence of such a tax and an accompanying BTA today is the same, although their motivation by environmental concerns is new.

This chapter makes two contributions.

First, we base our discussion on the model results in chapter 3, which establish that the introduction of an import tariff on the CO_2 content of the imported goods in the presence of a domestic carbon tax (i.e. the EU ETS) unambiguously lowers global emissions. A border tax can thus be motivated by appeal to global welfare and not by competitiveness concerns. This has important implications as we will show below.

Second, we examine the WTO compatibility of a concrete border measure, namely a tax on the content of CO_2 of imported goods without any rebate to exports (unlike most existing BTAs). This feature of our border measure makes it less 'protectionist' and hence more likely to be WTO-compliant. This is an important detail that is often overlooked in the existing literature on the WTO-compatibility of a carbon border tax adjustment. More often than not, the authors do not describe the precise design of a carbon border tax whose WTO-compatibility they discuss.[20]

---

[20] To the best of our knowledge, Genasci (2008) is the one exception.

Then we give some background information on the GATT Working Party definition of border tax adjustments (BTAs) and the WTO laws they are based on. After identifying the GATT rules that any border measure has to abide by, we finally discuss the design of a border measure that the EU should implement on imports before we conclude with policy recommendations.

4.1 BTAs in the WTO law: A review of the literature

The WTO rules on BTAs were set in the GATT Working Party report (1970) where a border tax adjustment is defined as "any fiscal measure which puts into effect, in whole or in part, the destination principle". Broadly speaking, governments can follow either of the following two principles. Following the 'origin principle' requires that the goods are taxed where they are produced, whereas in the case of the 'destination principle', goods are taxed where they are consumed. Only the 'destination principle' requires a border tax adjustment. This adjustment requires the refund of previously paid production taxes in return for an offsetting tariff on imports.

From an economist's point of view, it was long believed that both the origin and destination principle were trade-neutral (see Lockwood & Whally, 2008 for a recent discussion).[21] At the time of the discussion of the Working Party on Border Tax Adjustments, it was agreed that the existing WTO rules assured trade neutrality of BTAs and hence no adjustment was made to the rules. Hence the Working Party agreed that the relevant rules that should apply to BTAs were Articles II and III on the import side and Article XVI on the export side.

Concerning imports, Article II gives the right to impose a charge equal to an internal tax on an import, whereas Article III ensures that the importing country does not abuse this authority by charging an excessive sum. The second paragraph of Article III on National Treatment on Internal Taxation and Regulation summarises the purpose of this article: "The products of the territory of any contracting party imported into the territory of any other contracting party shall not be subject, directly or indirectly, to internal taxes or other internal charges of any kind in excess of those applied, directly or indirectly, to like domestic products."

---

[21] See Grossman (1980) for a discussion on trade-neutrality of uniform indirect taxes in the context of intermediate goods.

According to Article II on Schedules of Concessions: a charge equivalent to "an internal tax imposed consistently with the provisions of paragraph 2 of Article III in respect of the like domestic product or in respect of an article from which the imported product has been manufactured or produced in whole or in part" is authorised.

Concerning exports, Article XVI on Subsidies outlines that two countries trading have to "recognize that the granting by a contracting party of a subsidy on the export of any product may have harmful effects for other contracting parties, both importing and exporting, may cause undue disturbance to their normal commercial interests, and may hinder the achievement of the objectives of this Agreement". Hence, "contracting parties shall cease to grant either directly or indirectly any form of subsidy on the export of any product other than a primary product which subsidy results in the sale of such product for export at a price lower than the comparable price charged for the like product to buyers in the domestic market.

Other relevant articles include Article I, which is the Most-Favoured Nation clause, Article VI on Anti-dumping and Countervailing Duties and Article VII on Valuation for Customs Purposes. Among these, Article I will be discussed in detail later. Since concerned parties on the EU's competitiveness suggested the introduction of BTAs, several studies started to examine the WTO-compatibility of border measures. Genasci (2008) argues that the legality of border adjustments for energy taxes has been an unsettled question and the legal uncertainty multiplied when the concept is extended to an emissions trading system. He adds that "designing a mechanism to adjust the cost of emissions allowances upon export in a manner that adequately protects international competitiveness without resulting in illegal subsidies would be difficult". On the import side, it is also difficult to determine different emissions costs borne by the same industry in different countries. Another study by Biermann & Brohm (2005) concludes that despite ambiguity in both GATT rules and case law, if the EU had a carefully designed strategy for BTAs, under certain circumstances, these measures can be found to be compatible with world trade law. However, given the legal uncertainty, there is a high probability that affected members of the WTO would challenge such energy tax adjustments before the WTO dispute settlement mechanism. Biermann & Brohm (2005) use the US Superfund Tax case, the US Ozone Depleting Chemicals (ODC) tax and the Tuna/Dolphin and Shrimp/Turtle cases to make their arguments. Ismer & Neuhoff (2004) also show that an ETS-BTA can be compatible with WTO rules. To assure WTO compatibility, the BTA has to be simple to implement and for that it has to "focus on the CO_2 emissions caused by processed materials and a separate treatment of electric energy input to take account of regionally varying fuel mixes". Both Biermann & Brohm (2005) and Ismer & Neuhoff (2004) base their arguments on competitiveness and carbon leakage. Van Asselt & Biermann (2007), on the other hand, look at a range of proposed measures based on different criteria, including environmental effectiveness, the need to consider differentiated commitment, responsibilities and capabilities, etc. The authors conclude that among a wide range of options, border-cost adjustments on imports may be feasible, even though their legal and political feasibility is unclear. Quick (2008) argues that since the EU ETS is not a tax directly levied on domestic products, a border tax on imports or a BTA would be considered by the Appellate Body as 'naked' discrimination for protection of local products.

To the best of our knowledge, the WTO/UNEP report (2009) is the only study to reach similar conclusions as the present report: the WTO does not have any clear-cut laws and rules to assess the compatibility of a carbon-motivated border tax and hence several different rules have to be invoked. Nevertheless, it is possible to design such a carbon border tax. In the following section, we discuss in detail each relevant GATT article and case law to show how a carbon border tax can be designed to be WTO-compatible.

4.2 The relevant WTO regime for a EU carbon border tax

According to the Marrakesh Agreement, which established the WTO, member countries have to follow mutually supportive policies to ensure an open and non-discriminatory multilateral trading system and ensure protection of the environment and the promotion of sustainable development at the same time. As such, the WTO members are allowed to adopt trade-related measures aimed at protecting the environment. However, climate change is not part of the WTO's ongoing work programme and there are no specific WTO rules related to climate change. Although the GATT report draws out a definition and rules for application for BTAs, current WTO law is not clear when such measures are motivated by environmental protection. Both WTO Director-General Pascal Lamy and the Head of the Trade and Environment Committee of the WTO have stated that border measures and their WTO compatibility, per se, are not topics of internal discussion. Rather, the work of the Trade and Environment Committee focuses on two specific points where trade policy meets climate change: i) the impact of current trade patterns on climate change and ii) whether trade policy can be used to fight against climate change. In order to establish which articles would apply to a carbon border tax, one must start with the discussions of the GATT Working Party.

The GATT Working Party report (1970) indicates that the relevant articles for border tax adjustments (BTAs) can be separated into two parts:

For the import side of a BTA.

Article II. Schedules of Concessions and

Article III. National Treatment on Internal Taxation and Regulation.

For the export side of a BTA:

Article XVI. Subsidies apply.

The Working Party also included Articles I, VI and VII in their discussions. However, in the case of a carbon border tax, the relevant articles are Articles II and III from the import side, Article I. General Most Favoured Nation Treatment and Article XX. General Exceptions.

The WTO articles that would be applicable for the export side of a BTA are not relevant for our analysis as we consider only an import tax without an adjustment for exports. This is a key feature of the model introduced in chapter 3 which does not include an adjustment on the export side. In fact, it should be noted that technically our discussion in this chapter is on the WTO compatibility of a 'carbon border tax adjustment' rather than a 'carbon border tax', given we do not foresee adjustment on the export side.[22]

Each of the articles mentioned above (with the exception of Article XX) sets out the rules that a carbon border tax adjustment has to abide by in order to be WTO-compatible: Articles II and III define 'how' the carbon border tax has to be implemented, Article I defines 'who' the tax should be imposed on and Article XX defines 'when' a tax is justified.

---

[22] The WTO/UNEP (2009) report differentiates between a 'border tax' and a 'border tax adjustment' as follows:

> A 'border tax' is a tax (or customs duty) imposed on imported goods, while a 'border tax adjustment' is an adjustment of the taxes imposed domestically on products when the goods are imported.

According to Article II, a carbon border tax cannot violate the EU's WTO market access commitments. GATT Article II prohibits tariffs above a ceiling. However, if the EU can argue that the carbon border tax on the CO_2 content of imports is the 'same treatment' as an 'internal measure' (i.e. a border-enforced internal measure), it would not contravene its Article II obligations. Thus, the key is that the carbon border tax is going to be applied as part of a domestic carbon tax, namely the EU emissions trading system (ETS). Provided that the measure is not in violation of Articles I and III, it is thus likely to be WTO-compatible.

Even if a carbon border tax is considered a border enforcement of the EU ETS, it cannot treat imported products less favourably than 'like' domestic products. As a first step, one should clarify what is meant by 'like' products. According to the Appellate Body, two products are 'like' under Article III:4 if they are in a 'competitive relationship'. In this case, even if a basic industrial product is produced by two different production processes, one being more-carbon intensive, they are still considered 'like'. Hence, a carbon border tax, which is a tax on a 'process' not on a 'product', would violate this clause as it would treat two 'like' products differently. During the discussion of the GATT Working Party on BTAs, certain gray areas were left untouched. One of these sources of ambiguity was the definition of a 'like or similar product'. In the report, the Working Party decided that there has been considerable effort spent in the past to arrive at a common definition of a 'like or similar product', but due to a lack of improvement, it was decided that the term should be examined on a case-by-case basis. This, in the case of a carbon-motivated BTA, should present the greatest challenge.

Article III also determines that imported goods cannot be treated worse than domestic goods. In this case, the EU has to be careful not to impose a carbon border tax that exceeds the domestic carbon tax. Here we argue that the EU ETS acts like a tax. If one can argue that from an economic perspective, the costs and benefits of a carbon price is known, then there is little difference between a carbon tax and a cap-and-trade system (Bordoff, 2008). In that case, the tariff has to be equal to the price of carbon. The question that is often asked at this point is whether or not one should adjust (lower) the tariff for the free allowances that the sector covered by the ETS receives. In the recent energy and climate change package, it was agreed that only 50% of total allowances are to be auctioned in 2013. In reality, however, industry sectors such as steel, cement or refining get most allowances for free. We argue that this does not alter the equivalence of the allowances as a tax. Even if the allowances are received free, they can be sold for cash and thus there is an opportunity cost to using one to emit a tonne of carbon. Free allocation of allowances can thus be viewed as a lump-sum transfer coupled with a tax.

Genasci (2008) states that due to the novelty of emissions trading it is not clear whether emissions trading systems can be coupled with a border tax adjustment in a WTO-compatible manner. Free allocation of allowances further complicates the case. Genasci (2008) argues that the allowance expenses, the variable cost of such allowances over time and among firms and sectors taking into account the initial allocation of grandfathered allowances should be factored into the equation (i.e. the rate of carbon border tax). Paulwelyn (2007) also argues that allowances should become part of the calculation for the BTA as it is not the cost of purchasing the allowances that matters but the obligation to hold them (i.e. there is an opportunity cost to holding the allowances). Although these are reasonable arguments, there is another aspect of allowances that requires attention. In terms of the WTO compatibility of a border tax adjustment, it may be argued that the free allocation of allowances constitutes an illegal subsidy, or in WTO language 'an actionable subsidy' covered by the Subsidies and Countervailing Measures (SCM) Agreement. Under the SCM Agreement, a subsidy is defined as a financial contribution and/or a benefit conferred by a government to its domestic industries so that a given sector can develop with lower production costs (i.e. improve its competitiveness). Under Article III.1 of the SCM Agreement, government subsidies that are contingent on export performance or the use of domestic over imported products are prohibited. A subsidy is said to be 'actionable' if it is granted to certain enterprises only and if it causes injury to the domestic industry of foreign competitors (Article V of SCM). In practice it is difficult to prove injury.[23] In fact this may be why so far there have been no complaints reported to the WTO's Dispute Settlement Body on EU emissions allowances particularly, nor on subsidy schemes that are in place to reduce specific industries' greenhouse gas emissions. Another explanation may be the now-expired Article VIII.2c of the SCM Agreement, which allowed one-

[23] The WTO/UNEP report (2009) states that "even if free allocation of allowances are found to be actionable subsidies covered by the SCM Agreement, 'the adverse effects' would have to be demonstrated for action to be taken by another WTO member".

time subsidies to be introduced to offset increases in production costs of firms adjusting to new environmental regulations. The subsidy was limited to 20% of the adaptation costs incurred. The SCM is currently under negotiation as part of the Doha Round. If and when the Doha Round is finalised, this may be one crucial area where the WTO may not only contribute to combat climate change but also be made compatible with post-Kyoto commitments. However, the introduction of border measures will most likely be introduced by full auctioning as the current free allocation is transitional and related to fears of production leakage. Full auctioning can be introduced very rapidly.

4.2.1 *Establishing the carbon content*

As mentioned above, we argue that the appropriate carbon border tax adjustment should be based on the carbon content of the import. As such, we diverge from the mainstream literature which concentrates on ways to calculate the costs of the emissions trading system – including an adjustment for allowances – as mentioned above. According to Article III, any WTO-member has the right to impose a tax on imports that is equivalent to a 'internal charge'. However, this is a rather narrow perspective to take. We argue that if the carbon border tax adjustment is only to be on the imports from the ETS-covered sectors, the efficiency of the carbon border tax would be reduced dramatically in terms of its impact on global welfare. First, even though *de jure* ETS covers certain sectors, *de facto* coverage is in direct proportion with the level of vertical integration of ETS-covered sectors with other domestic sectors. In other words, the ETS has an impact also on those industries where e.g. steel or cement is used as an input. Second, every country uses a different production process, which determines the level of carbon emissions released into the atmosphere from an additional unit produced. Hence, the carbon content of a product is the appropriate base for a carbon border tax adjustment. To give an example, the CO_2 content in China's audio and video equipment exported to the US in 2003 was 27.4 $MtCO_2$, whereas for the same quantity of audio and video equipment, US exports emitted only 21.2 $MtCO_2$ (Bin & Harriss, 2008). The table below shows the difference in the CO_2 content of China's top five exports to the US versus the top five US exports to China. This difference is largely due to the relatively high use of coal and less efficient manufacturing technologies in China.

Table 4.1 Top five exports in US-China trade, ranked by their CO_2 embodiment in 2003 ($MtCO_2$)

CO_2 emissions embodied in Chinese exports		CO_2 emissions embodied in US exports	
Audio and video equipment	27.4	Soybeans	1.7
Games, toys	25.7	Plastics material and resin	1.1
Other computer peripheral equipment	19.4	All other basic organic chemicals	1.1
Institutional furniture	13.5	Fertilisers	0.8
Women's footwear	12.8	Aircraft	0.8
Total	98.9	Total	5.5

Source: Bin & Harriss (2008).

Since the carbon content of a product would be the tax base under a border tax, the 'process' of production rather than the 'product' itself becomes important. Related to the discussion on border measures being imposed on a 'process' and not on a 'like' product, WTO case law suggests that differentiation between 'like' products is not permitted under Article III but it can be under Article XX. Basing the tax on the carbon content of the product also makes sure that a border tax will be compatible with Article I. According to Article I, the EU can not single out countries that are not signatories to the Kyoto Protocol. In other words, the EU cannot discriminate against an import based on its 'origin'. In such a scenario, the EU may argue that the measure is not imposed based on origin but on process. If the EU decides to impose the carbon border tax on only certain countries, this is most likely to violate Article I as the MFN obligation is unconditional and broad. This could also raise complaints by the affected countries on the grounds that the EU's ETS covers only half the emissions. Indeed, from a WTO-compatibility perspective, the low coverage of the EU ETS will be problematic.

One crucial detail in the implementation of a carbon border tax adjustment is the methodology to be used for those countries that do not provide such information on their exports. Based on the US gasoline case, it is likely that the Appellate Body would not allow the EU to use one baseline for foreign producers and another for domestic producers. The best solution (under WTO compatibility considerations) would be to use an 'external' standard, such as ISO 14067, to calculate the carbon footprint

(CFP) of imports. As Radunsky & Laabs (2009) explain, the carbon footprint:

- refers to the calculation of the amount of GHG emissions associated with a company, event, activity, or the lifecycle of a product/service;
- enables ascertaining and managing the emissions of GHGs along the supply chain;
- allows a comparison of the difference of the CFP between different products (e.g. oranges and tomatoes), between the same product of different producers and between the same product of the same producer in different locations/shops;
- safeguards the survival of companies in the changing regulatory and economic business landscape;
- furthers the understanding of the risks and opportunities in the supply chain; and
- allows focusing of effort in response to new regulatory, shareholder and consumer pressures.

It thereby qualifies as a powerful tool to de-carbonise the supply chain of products.

As the authors argue, there are several initiatives underway (e.g. ISO 14067, Carbon Disclosure Project, PAS 2050 in the UK, Japan, New Zealand, Korea and California and recently, the EU ETS benchmarking exercise) to calculate or assess the life cycle of greenhouse gas emissions of goods and services. Among these, the ISO 14067 deserves special attention as this can provide the EU with the 'external' benchmark it needs to use when calculating the CFP of imports.

Many major multinational retailers have already launched bottom-up initiatives to measure a carbon footprint of their products (see Box 4.3 below).

Box 4.2 Standardisation at ISO

International standards help to assure quality, allow for comparability and contribute to market development. ISO standards are a powerful tool for developing global markets and supporting the harmonisation of government policies as well as supporting the acceptance of energy-efficient and low-carbon technologies and paving the way for the development and use of new technologies, including alternative renewable sources.

ISO/TC 207 "Environmental management" started a process on CFP in its meeting in Beijing (July 2007). An important milestone was the agreement on a new work item proposal (November 2008) related to the quantification and communication of the carbon footprint of products. The two-part standard ISO 14067 is developed in ISO/TC 207/SC 7/WG 2 "GHG in the value and supply chain". In developing ISO 14067, ISO/TC 207/SC 7/WG 2 can draw on a range of standards (e.g. ISO 14040, ISO 14044, ISO 14025, ISO 14064) and the experience in their application.

The development of ISO 14067 faces various challenges. The first is practicality. The standard needs to be practical as companies are choosing to monitor the CFP of a broad range of thousands of products, including their progress towards de-carbonisation on a yearly basis. In addition, there is the challenge to combine concepts and the terminology of a range of existing ISO standards that have been developed by different working groups. To agree on a meaningful communication of the CFP that addresses the needs of purchasers and consumers will be challenging.

The significant engagement, the increasing willingness to follow the process and to actively engage in ISO/TC 207/SC 7/WG 2 as well as the contribution of highly qualified experts with different backgrounds seem to offer a sound basis on which to manage those challenges. It is the common goal to make available ISO 14067 as a draft international standard by the beginning of 2010 and to issue ISO 14067 one year later by the end of 2011.

Source: Radunsky & Laabs (2009).

Box 4.3 Companies measuring GHG emissions in their supply chain

- Acer
- Banco Bradesco
- Boeing
- BT Group
- Cadbury
- Carrefour
- Celesc
- Colgate Palmolive
- Dell
- Exelon
- Fiji Water
- Heinz
- HP
- IBM
- Imperial Tobacco
- Johnson & Johnson
- Johnson Controls
- Juniper Networks
- Kellogg's
- L'Oréal
- Merrill Lynch & Co.
- National Grid
- Nestle
- Newmont Mining
- Pepsi Cola
- Procter & Gamble
- Prudential
- Reckitt Benckiser
- Royal Mail
- SSL International
- Tesco
- Unilever
- Vale
- Vodafone

Source: Radunsky & Laabs (2009).

Box 4.4 Benchmarking for the EU ETS

The revised EU Emissions Trading Scheme (European Parliament and Council of the EU, 2009a) foresees auctioning of allowances as the default option. However, Article 10a of the Directive allows for transitional free allocation to industry, based on Community-wide *ex-ante* benchmarks, which will be phased out for sectors not exposed to carbon leakage in steps by 2027 (80% in 2013, 30% in 2020). Installations in sectors or sub-sectors deemed to be "exposed to a significant risk of carbon leakage" are due to receive as of 2013 up to 2020 allowances free of charge at 100% of the benchmark.

The benchmarks will be based on a number of objectives, essentially related to providing incentives for GHG reductions and the use of energy-efficient techniques. The starting point is the "average performance of 10% most efficient installations" (in terms of GHG emissions) in a sector or sub-sector in the Community in the years 2007-08. Importantly, the farther an installation is from the most efficient one in the sector, the less likely it is to receive all of its allowances for free. The European Commission has launched a process to cope with benchmarking and has contracted a consultant - Ecofys, in cooperation with the Fraunhofer Institute for Systems and Innovation Research, and the Öko-Institute - to assist in this process.

Benchmarking offers the possibility to improve performance and inform EU and member state industrial policy (e.g. give indications of performance, potential, technological challenges, etc.). Benchmarking also offers possibilities at the international level. It may: i) become a useful tool to ease the transition to a low-carbon economy, ii) help industry to identify technology or performance opportunities and/or iii) provide information on best-practice. There might even be a possibility to 'transfer' the EU or parts of the EU benchmarking exercise internationally.

A list of 'exposed sectors', estimated to account for around a quarter of emissions covered by the EU ETS and around 75% of the emissions from manufacturing industry in the EU ETS, has been drawn up by the European Commission (2010a).

Box 4.5 Bottom-up approaches to carbon footprint labelling by Tesco

In the UK, the Carbon Trust, working with Defra (Department for Environment, Food and Rural Affairs, now DECC) and the British Standards Institute (BSI), has developed a standard methodology for measuring the greenhouse gas emissions from products and services, called Publicly Available Specification (PAS) 2050. It draws on existing best practice to create a single standard that will enable a consistent approach to measuring the embodied greenhouse gas emissions from products and services across their lifecycle, and is applicable to a wide range of sectors and product categories. It is anticipated that this work will be the first step towards an internationally agreed standard for measuring greenhouse gas emissions at the product level.

In 2007 the UK supermarket chain Tesco cooperated with the Carbon Trust, the environmental consultancy ERM and a small number of suppliers to measure the carbon in a range of products. The group used the methodology set out in the existing draft of PAS 2050 to measure the footprint of our products. This helped the Carbon Trust and BSI to test the draft standard and [its] work has helped inform the further refinement of PAS 2050.

During October, November and December 2007 [Tesco] measured the footprints in five product areas: laundry detergent, light bulbs, orange juice, potatoes and tomatoes. [Tesco] held workshops for [its] suppliers to explain life-cycle data and footprint calculation, consistent with the PAS 2050 standard so [Tesco] provided [its] suppliers with guidance documentation and a data collection template. [Tesco's] environmental consultants, ERM, supported [its] suppliers through the data collection and footprint calculation process, helping the suppliers with any questions they had. Once the results had been calculated the work was certified by the Carbon Trust.

[Tesco] managed to undertake this work with a fast turnaround. 30 products were measured and certified in 12 weeks. The following table shows the information presented on the CFP label.

Category and Product	Carbon footprint grams of CO_2e per functional unit (rounded)	Carbon footprint broken down into lifecycle stages (%)				
		Production	Distribution	Store	Use	End of life waste management
Detergents						
Tesco Non-Biological Liquid Capsules	700g per wash	17%	0.2%	1%	72%	10%
Tesco Super Conc. Non-Bio Liquid Wash	600g per wash	11%	0.1%	0%	83%	6%
Tesco Non-Biological Liquid Wash	700g per wash	17%	0.2%	1%	73%	9%
Tesco Non-Biological Tablets	850g per wash	32%	0.1%	0%	62%	5%
Tesco Non-Biological Powder	750g per wash	25%	0.1%	0%	69%	6%
Orange Juice						
Tesco 100% Pure Squeezed Orange Juice	360g per 250ml	91%	1%	7%	0.3%	1%
Tesco Pure Orange Juice From Concentrate	260g per 250ml	88%	2%	9%	0.5%	1%
Tesco Pure Orange Juice (1 litre)	240g per 250ml	93%	1%	4%	1%	1%
Tesco Pure Orange Juice (3x200ml)	220g per 250ml	93%	1%	5%	0.5%	1%
Light bulbs						
60W Pearl Light bulb	34kg per 1000 hrs of use	1%	<0.1%	<0.1%	99%	<0.1%
100W Pearl Light bulb	55kg per 1000 hrs of use	1%	<0.001	<0.001	99%	<0.1%
11W CFL	6.5kg per 1000 hrs of use	1%	<0.001	<0.1%	99%	<0.1%
20W CFL	12kg per 1000 hrs of use	2%	<0.001	<0.001	98%	<0.1%
11W Spotlight	6.5kg per 1000 hrs of use	2%	<0.001	<0.1%	98%	<0.1%
60W Spotlight	34kg per 1000 hrs of use	1%	<0.1%	<0.001	99%	<0.001
Potatoes						
King Edwards (2.5 kg)	160g per 250g serving	33%	1%	3%	56%	7%
Anglian New (2.5 kg)	140g per 250g serving	34%	1%	3%	58%	4%
Organic New (1.5 kg)	160g per 250g serving	40%	1%	4%	51%	4%
Organic Baby New (750 g)	140g per 250g serving	48%	1%	5%	41%	4%

Source: TESCO website (http://www.tesco.com).

After a decision has been made on which outside benchmark should be used to calculate imports' carbon content if they are not already provided by the producer, the next step is implementation. Although calculating the carbon content of thousands of products from several different countries may seem a daunting job, one could argue that it should be no more complicated than obtaining a CE marking, for example, to enter the EU market.

The CE marking is a mandatory conformity mark on several different products that producers are required to display if the product is to be sold in the European Economic Area, or EEA (consisting of three of the four EFTA member states – Iceland, Liechtenstein and Norway – the EU-27 and the European Community). The CE marking certifies that the product has met EU consumer safety, health or environmental requirements. Certain aspects of how CE marking is implemented are also appealing for enforcement of displaying the carbon content of a product. For example, if the producer can identify the directives his product has to be in conformity with, he can himself declare the conformity sign. Otherwise, there are several consultants who can visit the location of production and certify the 'process' of production (rather than the product) so that the manufacturer does not have to obtain a CE marking each time it ships the same product to the EU and the EEA market. This can be directly applied in the case of a carbon content of a product.

4.2.2 *Article XX*

The key element of the carbon border tax that we are discussing in this paper is that the tax has to be levied on the carbon content of an imported product; as such, it is not a tax on a 'product' but a 'process'. As discussed above, one of the most important questions in the case of a carbon-motivated tax is to establish whether or not products that are produced using different production methods can be considered 'unlike'. The WTO claims: "When comparing two products, different processes or production methods (PPMs) used in the manufacture of such products do not per se render these products 'unlike'". In addition, the WTO rules are clear on 'product' taxes but not on 'process' taxes. Hence the legality of BTAs on 'process' taxes is unclear. In the case of a carbon tax of the type considered here, the tax is primarily going to be a tax on a 'process' and not on a 'product'. One way of justifying a 'process' method is by using the exemptions provided under Article XX.

We showed in chapter 3 that border measures are effective tools to fight climate change (i.e. reduce global CO_2 emissions and increase global welfare). Moreover, one can also argue that a low-carbon atmosphere is an 'exhaustible natural resource' (Article XX(g)) and that numerous international agreements have recognised the importance of combating climate change.

Article XX on General Exceptions allows trade restrictions if they are necessary to achieve an environmental goal.[24] This is crucial because it signifies that a carbon border tax whose aim is to increase global welfare can be WTO-compatible, but not carbon-related border measures motivated by competitiveness concerns.

Our motivation for a carbon border tax is also different from the concerns over 'carbon leakage'. Several studies indicate that carbon leakage in the case of no-border measures is rather small (e.g. Veenendaal & Manders, 2008). This can be explained by the fact that according to recent figures (European Environment Agency, 2007), 44.1% of the CO_2 [25] emissions in 2005 by the EU-27 were accounted for by the non-traded sector: 27.4% by transport and 16.8% households and services.[26] The figures below indicate that while the contribution of the energy industries and industry to CO_2 emissions has declined slightly over the last decade, the contribution from transport has increased considerably.

---

[24] As emphasised in chapter 3, any economic analysis of carbon border measures should not only consider CO_2 emissions but also the cost of production and consumer welfare. Our analysis concludes that an import tariff improves global welfare (taking into account also producers and consumer surpluses), as such a measure partially imposes carbon pricing in those countries where there are no domestic measures.

[25] In this study we strictly refer to CO_2 emissions, as the ETS mainly covers CO_2 although greenhouse gas emissions include six gases. The ETS coverage of GHG however is likely to change.

[26] Despite the heavy transport costs for the trade in energy-sector products, some concerns have been raised within the EU that production facilities may be exported to neighbouring countries, e.g. Ukraine.

Table 4.2 CO_2 emissions[a] by sector, EU-27 (shares, %)

	1995	2005
Energy industries	**35.2**	**34.3**
Industry	**23.1**	**20.9**
Transport	**23.3**	**27.4**
- Road transport	75.5	71.9
- Civil aviation[b]	9.9	12.0
- Navigation[b]	12.7	14.7
- Railways	1.0	0.6
- Other transport[c]	0.9	0.8
Households, services, etc.	**17.6**	**16.8**
- Households	11.2	10.5
- Services, etc.	6.4	6.3
Other[d]	**0.8**	**0.6**
Total	**100.0**	**100.0**

[a] Including international bunkers but excluding LULUCF (land use, land-use change and forestry) emissions.

[b] Including international bunkers (international traffic departing from the EU).

[c] Includes pipeline transportation and ground activities in airports and harbours.

[d] Solvent use, fugitive, waste, agriculture.

Source: European Environment Agency (2007).

It has to be noted that the EU ETS does not currently cover the transport sector or residential housing. From a legal point of view, it can be argued that the measure (i.e. the carbon border tax) is not an effective tool to serve its purpose (i.e. reduce CO_2 emissions). If the domestic or 'internal' measure (i.e. the EU ETS) does not address fully or only covers partially the major contributing sectors to the problem, the EU is more likely to face challenges on any border measure it imposes on other countries.

The European Commission is contemplating the introduction of EU-wide carbon taxation in the non-ETS sectors. Such a tax would add explicit CO_2 tax components to the 2003 Directive on energy taxation. Directive 2003/96/EC sets minimum tax rates in the member states for the use of various fossil-fuel types and of electricity, with no upper limit (Council of the EU, 2003). Using rates that have been circulating in early European

Commission drafts of €0.01/kg CO_2 for heating fuels and either €0.01 or €0.03 per kg CO_2 for motor fuels,[27] depending on their use, would impose CO_2 cost on the non-ETS equivalent to an emissions price of €10 per tonne (and maybe €30 per tonne for motor fuels) which would then affect directly the entire EU economy.

Moreover, the idea of common carbon taxation is starting to attract the support of member states as an uncomplicated and cost-effective market-based tool for reaching the new non-ETS targets in the climate and energy package, for motivating energy efficiency and for raising proceeds, while avoiding internal market distortions in case of unilateral taxation in some countries only. But the unanimity in the Council of the EU required for passing tax legislation may be hard to achieve due to national budget peculiarities and concerns over imposing a disproportionate burden in poorer countries. But the proposal could be adopted under the 'enhanced cooperation' clause, which would allow a majority of member states to adopt such a law. From the very beginning, the Swedish Presidency made clear its support for carbon tax (see Reinfeldt, 2009). The Council of the EU and the European Parliament had also previously encouraged carbon taxation by stressing the importance of introducing market-based instruments, such as taxes, which would be separate from the EU ETS permits (Council of the EU, 2007; European Parliament, 2008). Individual approaches have been developed by some member states. Carbon taxes with generally higher rates than the newly proposed minimum already exist in Sweden and Finland and have been recently proposed in France (IEA/OECD, 2009; Sarkozy, 2009).

4.3 Case law

There is now some evidence that although the WTO rules have not changed, the panel rulings on trade and environment cases have started to recognise that trade policy can be used to protect the environment. For example, the US action against shrimp catchers provides an interesting example of justifiable discrimination between products on the basis of PPMs to achieve an environmental goal. The case involved the imposition of an import ban by the US on shrimp that was harvested by a method that

---

[27] Motor fuels from waste and other residues have been added to the Directive for the first time, but are not subject to CO_2 tax.

led to incidental killing of sea turtles. The exporters were exempt from the ban if they proved that they caught shrimp by the use of a certain piece of equipment that did not threaten sea turtles.

The EU carbon border taxes can also benefit from the US experience with environmental taxation. The Superfund Chemical Excises (1986) and the Ozone-Depleting Chemicals (ODC) taxes are two such examples with 'process' BTAs. The purpose of the Superfund tax was to raise revenue to clean up the contaminated toxic waste sites, whereas the purpose of the ODC tax was to discourage the use of chlorofluorocarbon (CFCs) and hydrochlorofluorocarbons (HCFCs) by increasing the price of taxed chemicals and discouraging their production. Both of these systems of taxes have characteristics that may be applicable in an EU carbon-tax system.

Most importantly, the GATT has approved the system of BTAs on process taxes used by the Superfund tax.[28] In response to a request from Canada, Mexico and the European Union, a GATT conciliation panel was formed to examine the consistency of the system of BTAs under the Superfund chemical tax with international trade rules. The GATT Superfund Panel Report found that for purposes of Article III (national treatment of 'like products) the tax on imports of taxable substances manufactured with taxable chemicals did not treat those goods differently than similar goods produced in the United States. The panel found that the tax was imposed on imported substances because they were produced from chemicals subject to an excise tax in the United States and the tax rate was determined in principle in relation to the amount of those chemicals used and not in relation to the value of the imported substance. The conciliation panel approved the system of BTAs on taxable substances based on the actual consumption of taxable chemicals in their production, and the system of imputation based on the predominant method of manufacture. However, it rejected the fallback BTA of 5% of the value on the grounds that it imposed a higher tax on imports than on similar domestic production.

---

[28] GATT Panel Report, *United States - Taxes on Petroleum and Certain Imported Substances*, L/6175, BISD34S/136, 154 ff., adopted on 17 June 1987.

4.4 Key elements for the design of a WTO-compatible EU carbon tax

Based on the above arguments, a WTO-compatible EU carbon tax should have the following main features:

- The EU carbon border tax should be a tax on all imports[29] and the tax base should be the carbon content of the imported product. This already introduces two important characteristics of a compliant border measure: first, it is a tax on a 'process' and not on a 'product'; second, as this is a tax based on a 'process', it is origin-neutral, and hence it does not violate Article I (MFN treatment).
- Since taxes based on 'process' violate Article III (national treatment of 'like' products), the EU has to invoke Article XX (general exceptions). Under Article XX, trade-related measures are allowed to protect human, animal or plant life or health. Therefore, the EU can argue that the import border tax (as per chapter 3) can increase global welfare by directly curbing CO_2 emissions globally. More specifically, the EU can argue that *a low-carbon atmosphere, necessary to avoid catastrophic climate change, should be viewed as an "exhaustible natural resource"* (as per Article XX (g)) (Bordoff, 2008). The panel outcome on the US gasoline case states that clean air is a resource susceptible to depletion.
- The rate of tax should be set equal to the difference between the EU price of carbon (ε) and the foreign price of carbon. Hence, the tax on a product is going to be equal to the carbon content times the difference in the EU carbon price and foreign carbon price.
- The border tax should tax the manufacturer or the importer at the point of the first sale or use of the imported product. The border tax should apply to all products as all products emit carbon during the production process. However, to reduce the administrative burden of the border measure, a threshold level for carbon-content of the product may be introduced.

---

[29] Some countries may challenge the EU by arguing that the EU ETS now covers only 50% of the CO_2 emissions. However, the EU can argue that other initiatives de facto provide a full coverage.

- The carbon tax has to apply to the *first domestic sale and use* of the 'product' (referred to as product Q in the model) which produces a negative externality. As such, the EU carbon tax should be a carbon tax collected at the importer level and therefore the EU carbon tax is subject to a border measure.
- The carbon tax should be revenue-neutral: revenues collected from the tax should be used to create a fund that will be used to ease the transition of energy-intensive industries and to invest in new technologies.
- The tax should be collected at first domestic sale or use, without a refund on exports.

The tax on an imported product should be determined in the following manner in order to ensure WTO compatibility:

- For those imports where the importer provides the carbon content of the product or the carbon footprint (using internationally accepted methodologies such as the ISO 14067), the import tax will be equivalent to the carbon content times the difference in price of carbon.
- For those imports where the importer does not provide the information on the carbon content, the carbon content of the product will be calculated using the ISO 14067 methodology.
- Finally, the EU has to take into account 'different conditions' in different countries. As Bordoff (2008) argues, failure to do so may violate the non-discrimination clause of Article XX chapeau.

The Appellate Body's interpretation of Article XX chapeau also suggests that the EU has to take the initiative to negotiate with other countries that might be affected by the EU's carbon border tax. This was suggested in the US shrimp case.

4.5 Conclusion

This chapter analysed the WTO compatibility of a border tax on the carbon content of imports to the EU. The motivation for carbon border taxes was presented in chapter 3, which indicated that a carbon border tax and an associated domestic carbon tax are welfare-increasing for the world. Based on the theoretical model, we argue that our motivation for the introduction of a carbon border tax is not based on competitiveness and carbon-leakage concerns of the EU industry as mentioned in other studies. This has implications from an international law perspective, as the model shows that indeed the measure (border tax) serves its purpose (to combat climate change).

We examined the current GATT rules and WTO case law to determine the appropriate design of a potential EU carbon border tax. The complexity of this legal case makes it difficult to determine whether such a border measure would be WTO-compliant. As discussed above, there are several articles of GATT that one has to take into account to avoid being found in violation of GATT rules. Although there are no rulings on carbon-motivated border measures, WTO case law suggests that the WTO has been adjusting to rising demands from the climate change community to consider using trade measures to aid climate change mitigation.

We show that carbon border tax (with no export refund) can be WTO-compatible if it does not discriminate between the foreign product and the domestic product and if it treats all WTO members equally. The key condition is that WTO rules (Article XX) provide certain exceptions to the general prohibition to impose import taxes. One admissible justification is that the measure aims to protect 'an exhaustible natural resource'. Since the stock of CO_2 in the atmosphere is such an 'exhaustible natural resource', trade-related carbon measures can invoke this exemption as a justification.[30]

One key difficulty the EU might face if it came to a WTO panel ruling on this issue is that the ETS does not 'substantially' cover all domestic sources of GHG emissions. However, the EU could argue that it has taken a commitment on *total* CO_2 emissions (which account for most of all GHG emissions in the EU) and that its policies thus cover substantially all sources of emissions.

A further key aspect of invoking the exemption under Article XX of the WTO is that the EU must make a good faith effort to negotiate other solutions with the contracting parties (which include all major emerging markets except Russia) before proceeding to impose border measures.

---

[30] This interpretation would seem to be strengthened by the fact that the United Nations Framework Convention on Climate Change (UNFCCC), whose aim is to combat climate change, has been ratified by almost all countries. Many statements from all global fora have also repeatedly documented the consensus that fighting climate change is an important global policy objective.

Box 4.6 Superfund chemical excises

The United States Superfund Amendments and Reauthorization Act of 1986 created a system of taxes to fund the clean-up of toxic waste disposal sites, including a petroleum products excise, a corporate income tax surcharge and a system of excises on taxable chemicals and substances. The main features of the Superfund tax scheme are as follows:

- The Superfund chemical excises apply to the sale or use of the enumerated chemicals in the US. As a result, the taxes are subject to BTAs appropriate to a tax on consumption collected at the level of the manufacturer. Imports are taxed on the first sale or use by the importer, and any tax previously collected on exports is rebated. Because the tax is collected on the first domestic sale or use, a rebate of tax on export is not always necessary.
- The BTAs apply not only to taxable chemicals enumerated in 26 USCA 4661, but also to untaxed chemicals manufactured using taxed chemicals as a feedstock. These chemicals are referred to in the tax code as 'taxable substances'. The Superfund tax applies no domestic tax to taxable substances. BTAs were created on taxable substances equal to the tax paid on the taxable chemicals used to manufacture those substances.
- Taxable substances are of two types: 1) substances on an initial list contained in the statute and 2) substances approved by the Secretary of the Treasury under an application process created by the statute. To be approved as a taxable substance, the taxpayer must establish that taxable chemicals constitute at least 50% of the chemicals used to produce that substance, by either weight or value.

Import-side: When a taxable substance is imported, a tax is imposed on the importer at the first sale or use. There is a three-tier system for determining the tax rate on an imported taxable substance.

- First, the importer may provide detailed information on the taxable chemicals actually used in the manufacture of the taxable substance. In this case, the tax is based on the amount of tax that would have been paid on the taxable chemicals if the taxable substance had been manufactured in the United States.
- Where the importer fails to provide such information, the Superfund legislation created two alternative systems for calculating tax liability. The US Treasury Department issues regulations stating for each taxable substance the amount of taxed chemicals used to produce that substance in the US under the dominant method of manufacture. Imports are then taxed based on the amount of taxable chemicals that would have been used to produce the good in the US using the main production method.

- Finally, where no regulation has been issued, a penalty tax of 5% of the value of the import was imposed.

The measure of the amount of tax on taxable substances for BTA purposes is the tax on the materials used in the manufacture of the taxable substance. This tax is not pro-rated by mass, weight or value when only a portion of the taxable chemical is incorporated into the taxable substance, provided the taxable chemical has been consumed in the manufacturing process. Thus, the Superfund taxable substance BTAs are BTAs on the manufacturing processes, not on physical products.

Source: Hoerner (1998).

5. The Criteria of the UN Framework Convention for Climate Change

The previous chapter elaborated on the compatibility of the border measures with WTO rules and found that a border tax might be imposed if it is equal to an internal charge and is necessary to protect an exhaustible resource. The first condition seems to be easily met. The second condition requires some interpretation of what is meant by the exemptions provided in Article XX. We argued that any interpretation of this article must be guided by the global consensus that fighting climate change is an important global policy objective.

A key aspect here is that the UNFCCC lays down the principle of "common but differentiated responsibilities and respective capabilities" (Article 4.1). This means that developed countries have more responsibility than developing countries and should take a lead in climate action. Developing countries argue that the differentiation part of the principle only applies to the relations between developed and developing countries. This argument is based on the current structure of the Convention dividing countries into two groups (i.e. Annex I - developed countries plus economies in transition - and non-Annex I countries). Some countries argue for a revision of the structure to more accurately reflect the current conditions for a post-2012 regime.

Nevertheless, this principle does not prevent countries from taking their own actions, regardless of the classification. The Convention also declares that both developed and developing countries make general commitments including the adoption of national policies and measures on mitigation (Article 4.2(a)). Already not only developed countries but also developing countries are implementing domestic policies and measures for mitigation. By imposing costs on industry and consumers, these actions create a shadow carbon price. In a world without a global cap-and-trade or a common indicator acceptable to emerging economies, such a carbon price could play an important role in comparison to efforts to measure their equivalence.

5.1 Key indicators for climate change

The UNFCCC principle introduces two key criteria that guide countries' commitments to climate actions: responsibility and capability. There are other equally important criteria (e.g. equity and simplicity). This report however concentrates on the UNFCCC principles. The responsibility criterion can be translated into several key pieces of data and indicators. The most essential data for climate change are *emissions* (e.g. all GHGs or CO_2 only; historical, actual or projected), *population* and *production* (e.g. GDP). Key indicators are derived from a combination of these data, e.g. per capita emissions, GHG or CO_2 intensity (emissions per unit of GDP).

The simplest way to translate the capability criterion would be to consider *wealth* (e.g. GDP or GDP per capita). The UNDP has developed composite indices, Human Development Indices (HDI).[31] The Millennium Development Goals (MDGs) also specify a number of poverty indicators.[32]

We examine below three examples - historical emissions, per capita emissions and CO_2 intensity - that highlight different types of difficulties associated with the choice of an indicator: for historical emissions, the retrospective identification of a GHG as a pollutant; for per capita emissions, the risk of creating a bias towards certain economic and social models; and for CO_2 intensity, environmental uncertainty.

Comparing responsibility and targets expressed with different or even with the same indicators is additionally hampered by data inconsistencies. Various sources base their indicators on different concepts in terms of which gases are considered (CO_2 versus all GHGs, etc.), and which emitting sources (industrial, land use, etc.) to include. The bulk of

---

[31] See the website of the UN Development programme (http://hdr.undp.org/en/statistics/).

[32] On the MDGs, see the World Bank's website (http://go.worldbank.org/EF35B3YI70).

the literature is fuzzy on these questions, resulting in many different claims and counterclaims. Not surprisingly, most countries emphasise the type of emissions data that best supports their (perceived) national interests.

CO_2 is often used as a proxy for total GHG emissions, as it is the most important greenhouse gas and holds the largest share of GHG emissions (see Box 1.1). However, whether CO_2 emissions from LULUCF are omitted or not makes a substantial difference (see e.g. Figure 5.1 and Figure 1.1). It must always be kept in mind that the shares of emissions from certain gases, e.g. CO_2, and sources, e.g. LULUCF or fossil fuel use, are different for each country and do not coincide with those shown in Figure 1.1 for the whole world.

Additional methodological problems include the lack of accurate data collection in many countries and differences in approaches to computing emissions of the same type. These often lead to discrepancies between database sources and have spurred a call for stricter and transparent emissions accounting rules (see, for example, European Commission, 2010b). The Kyoto Protocol required only Annex I countries to conform to the common methodology and metric in their 'national communications and inventories'. These data issues are usually denoted under the rubric 'monitoring, reporting and verification (MRV)'. Among Annex I countries MRV is no longer an issue. However, the major non-Annex I countries have so far not subscribed to MRV obligations. The accuracy of data from these countries might thus be subject to doubts and discussions for some time to come.

Since data on CO_2 emissions from fossil fuel combustion are the most widely available, we have therefore selected them for the comparisons in Figures 5.1 and 5.2. CO_2 emissions by the IEA do not include those from industrial processes that do not involve energy-related uses of fossil fuels, such as cement production (see Figure 1.1), which constitutes approximately 4.5% of CO_2 emissions in the World in 2006, but 3% in the EU and 9.6% in China.[33]

---

[33] Calculations based on data from CAIT 7.0, (Climate Analysis Indicators Tool), a comprehensive and comparable database of greenhouse gas emissions data (including all major sources and sinks) and other climate-relevant indicators developed by the World Resources Institute (WRI) (see http://cait.wri.org/).

5.1.1 *The stock of historical emissions*

Historical emissions matter because climate change is a stock problem. Climate impacts are caused by concentrations, not emissions of GHGs in the atmosphere. GHGs typically stay in the atmosphere for 100 years. This means, for example, that historical emissions for up to the last 100 years are still responsible for causing climate change. Countries that started industrialisation earlier (e.g. the UK) have contributed a higher share of GHG emissions to the global GHG stock. On the other hand, GHG emissions at the time were not recognised as being a problem. Responsibility also changes when land-use change is included in the overall calculations.

Countries with more recent industrialisation have a tendency to emphasise historical emissions, but this is already changing in countries, such as China, where the growth of emissions is so rapid that their historical emissions will become relatively large within the near future.

5.1.2 *Per capita emissions*

A country's per capita emissions reflect mainly its level of income per capita, but the growth of per capita emissions depends crucially on its socio-economic development model. Developing countries follow a wide range of economic growth models. While China's manufacturing goods export model is based on high-carbon emissions, this is far less the case for India, where the service sectors are relatively more important. This is why the CO_2 intensity of India's economy is about one third lower that of China (see also Table 5.2 below). The more important differences among emerging economies are, however, between the those countries with an important industrial sector (e.g. China, India and Indonesia), whose CO_2 intensities of GDP are much higher than those of most Annex I countries and others, like Brazil and Mexico, which are much closer to the bottom range of Annex I countries (however, this difference might be due to LULUCF, as explained above). As the focus of international negotiations has shifted to CO_2 intensity, we discuss this metric separately below.

Wide differences exist equally among Annex I countries. For example, the US and Canada emit 100% more per capita (and GDP) than EIT Annex I or Europe Annex I (see Figure 5.1).

Figure 5.1 Annual per capita emissions, by country, 2005 (tCO_2e)

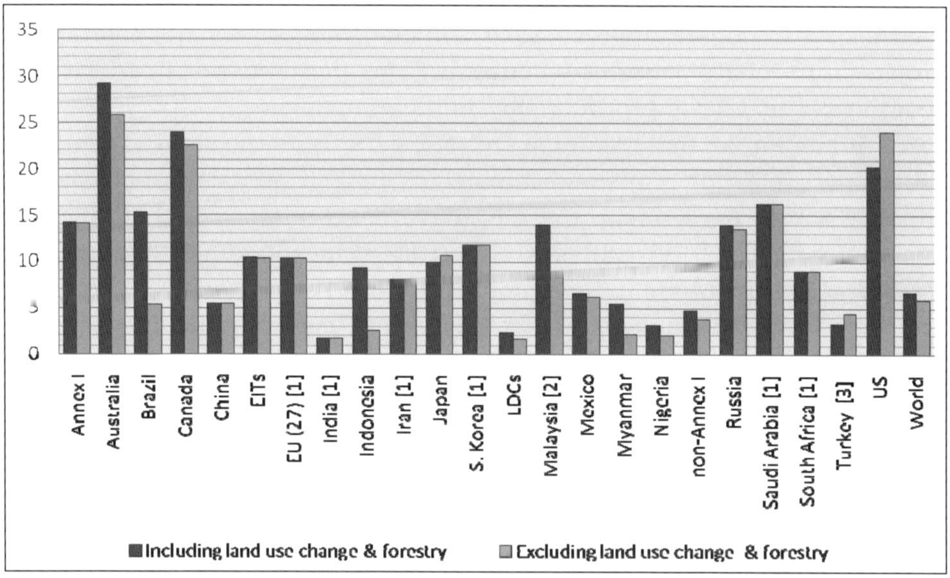

[1] Data from Land Use Change & Forestry not available.

[2] PFC, HFC & SF6 data not available.

[3] PFC data not available.

Source: Based on data from Climate Analysis Indicators Tool (CAIT) Versions 7.0 and 3.0, developed by the World Resources Institute, Washington, D.C.

Most of the differences in the figure can be explained by 'objective' conditions such as GDP per capita, geography, power sector structure as a result of fuel availability and the structure of the economy or industry in question. Others relate to lifestyle and reflect societal preferences.

It is clear that relatively poor countries tend to emphasise per capita emissions in international negotiations.

5.1.3 *CO_2 intensity of the economy*

CO_2 intensity is usually defined as CO_2 emissions[34] per unit of GDP. This measure is of course favoured by fast-growing economies, but it also has

---

[34] CO_2 intensity of the economy is usually taken as a proxy for GHG intensity (which includes both CO_2 and other GHGs). As a proportion of their GHG intensities, however, the CO_2 intensities of various countries differ, because the shares of emitted GHGs are not the same as for the world total (see Fig. 1.1). Both

the advantage that it should be closely related to the economic cost of achieving certain targets. It is clear that the economic cost of keeping emissions merely constant will be much higher for an economy that grows at 9% per annum than for one that does not grow at all.

For example, the EU's target of reducing emissions[35] by about 13% over the horizon 2005-20[36] translates into an implied reduction of CO_2 intensity of roughly 2.4% per annum (or -30.7% over the whole period). This is the result of combining the fall in emissions (-0.9% p.a.) with a potential GDP growth rate of around 1.5% per annum on average.[37]

For comparison, the pledge submitted by the US in the context of the Copenhagen Accord (emissions reduction "in the range of" 17% in 2020 from 2005 levels[38]) would amount to a CO_2 intensity reduction of about 3.2% per annum on average (or -38.3% for the period).[39] This is because the potential GDP growth rate of the US is estimated to about 0.5% higher than that of the EU, i.e. at 2% per annum on average. Higher US growth rates would of course have to result in higher reductions of US CO_2 intensity if the US is to attain its target. If both the EU and the US were to attain their targets, the outcome would probably be quite comparable (in terms of intensity changes), taking 1990 as the base year, because over that longer

---

the CO_2 intensity and the GHG intensity of the economy usually exclude CO_2 emissions from LULUCF.

[35] We assume a constant portion of CO_2 emissions in total absolute GHG emissions. In this case the percentage changes presented in this section are valid for both total GHG emissions (absolute targets are announced in this metric) and for the CO_2 emissions (intensity targets have been announced in terms of CO_2).

[36] The target of -20% from 1990 levels corresponds to -13.1% from 2005 levels (own calculations) due to the 7.9% reduction (EEA, 2009) achieved by the EU until 2005.

[37] Per annum averages are compound annual growth rates. GDP growth rates are own estimates based on the data and projections until 2014 by the IMF (2009).

[38] The US letter to the UNFCCC Secretariat can be found on the UNFCCC website (see "Appendix I - Quantified economy-wide emissions targets for 2020" at http://unfccc.int).

[39] For comparison, see the corresponding World Resources Institute's estimate of US (-37%) and EU (-30%) efforts in terms of GHG intensity improvement (including LULUCF) from a 2005 base year under slow growth (Levin & Bradley, 2010).

30-year period, the difference in the change in emission levels (-20% for the EU, versus only -3% for the US) would be offset by higher US growth.

One could thus argue that the US effort (at least going forward from 2005) is more ambitious than that of the EU – at least in the sense that the improvement in intensity terms would have to be about 0.8 percentage points higher, compensating for the slightly steeper improvement by the EU since 1990 (see Figure 5.2).

Historically OECD economies have been hard-pressed to improve energy (and emissions) intensity by more than 1% per annum. Both the EU's and the US 2020 targets would thus require considerable efforts.

Figure 5.2 CO_2 intensity trends in selected economies, 1990-2006 (percentage growth from 1990)

Source: Based on data from Climate Analysis Indicators Tool (CAIT) Version 7.0. World Resources Institute, Washington, D.C., 2010.

Among mature economies, CO_2 intensity might be a useful measure. But is it not clear whether one can compare this measure across emerging market economies whose CO_2 intensity is currently much higher and accompanied by the potential for savings.

For example, if China were to continue to grow at 9% per annum but to achieve only the pace of improvement in CO_2 intensity (implicitly) planned by the EU of about -2.4% per annum, its emissions levels would continue to increase by more than 6% per annum. This would imply that over the 15 years between 2005 and 2020, China's emissions would exceed its growth by more than 150% (and more than they have done over the last 15 years). By 2020, China would then emit more than the EU, the US and Japan combined!

The pledge recently offered by the Chinese government, in the context of the Copenhagen Accord, is closer to the intensity improvement implicitly promised by the US, as calculated above (-38%) (and would practically match it if the US potential growth post-crisis were higher).[40] This Chinese target is to reduce emissions intensity (CO_2 emissions/unit of GDP) by 40-45% between 2005 and 2020,[41] corresponding to between -3.3% and -3.9% annually on average over 15 years. At the 9% annual growth in the example above, this would translate into more than a doubling of 2005 absolute emissions (increase between 100% and 119%).

In order to be comparable, commitments by fast-growing emerging market economies must thus necessarily be much more ambitious in terms of GDP intensity than those of mature economies if the 2° C target is to be reached. This makes sense if one considers that the abatement costs will be lower the higher the growth rate, and thus the rate of renewal of the capital stock.

---

[40] The World Resources Institute has estimated the US intensity improvement at 43% under higher GDP growth and 37% under lower growth, under its target of 17% absolute reduction (Levin & Bradley, 2010).

[41] The official letter to the UNFCCC Secretariat reads that "China will endeavour to lower its carbon dioxide emissions per unit of GDP by 40-45% by 2020 compared to the 2005 level" (see "Appendix II - Nationally appropriate mitigation actions of developing country Parties" on the UNFCCC's website at http://unfccc.int).

Table 5.1 Key statistics on selected Annex I parties

Energy-related[a] CO_2 emissions	**Tonne CO_2 /capita 2007[b]**	**CO_2 intensity energy mix 2007[c]**	**CO_2 intensity of GDP 2007[d]**
EU-27	7.9	53.3	0.4
Australia	18.8	76.3	0.78
Canada	17.4	50.8	0.66
Japan	9.7	57.5	0.24
Russia	11.2	56.4	0.91
Turkey	3.6	63.3	0.71
Ukraine	6.8	54.6	6.01
US	19.1	58.9	0.5

[a] Total CO_2 emissions from fuel combustion using the IPCC Sectoral Approach; excluding international bunkers and aviation.

[b] Data from IEA (2009a).

[c] Tonne CO_2/terajoule of total primary energy supply; data from IEA (2009a).

[d] kg CO_2/$ using 2000 prices and exchange rates; data from IEA (2009a).

Source: Data from IEA (2009a), table format adapted from European Commission (2009d, p. 23).

The UNFCCC's crude categorisation of Annex I and non-Annex I countries hides huge and growing differences among the latter group of developing and emerging countries. India is a fast-growing emerging economy but has very low per capita emissions (a little more than 1 tonne of CO_2 per capita compared with about 4.6 tonnes of CO_2 per capita for China, measured in 2007) comparable to a least developed country (LDC). Table 5.2 summarises the main data of major economies including emerging economies and shows a significant variance among them. Note that the figures for Brazil and Indonesia would be much higher if the impact of land use, land-use change and forestry (LULUCF) were included.

Table 5.2 Key statistics on carbon emissions in emerging economies

Energy-related[a] CO_2 emissions	**GDP/ cap 2007[b]**	**Tonne CO_2/ Capita 2007[c]**	**CO_2 intensity energy mix 2007[d]**	**CO_2 intensity GDP 2007[e]**	**CO_2 emissions 1990[f]**	**CO_2 emissions 2007[f]**	**Change CO_2 1990-2007[f]**
Brazil	9,567	1.8	35.2	0.4	193	347.1	79.8%
China	5,383	4.6	73.6	2.3	2244	6071	170.6%
India	2,753	1.2	53.2	1.7	589.3	1324	124.7%
Indonesia	3,712	1.7	47.3	1.6	140.2	377.2	169%
Mexico	14,104	4.1	56.8	0.6	292.9	437.9	49.5%
South Africa	9,757	7.3	61.5	1.9	254.7	345.8	35.8%
South Korea	24,801	10.1	52.5	0.7	229.3	488.7	113.1%

[a] Total CO_2 emissions from fuel combustion using the IPCC Sectoral Approach; excluding international bunkers and aviation.

[b] US$ per capita in purchasing power parity terms (PPP); data for 2007 from UNDP (2009).

[c] Data from IEA (2009a).

[d] Tonne CO_2 / terajoule; data from IEA (2009a).

[d] kg CO_2 / US$ using 2000 prices and exchange rates; data from IEA (2009a).

[f] Million tonnes of CO_2 ($MtCO_2$); data from IEA (2009a).

Source: Data from IEA (2009a), UNDP (2009), table format adapted from European Commission (2009d, p. 24).

5.2 Mid- and long-term global emissions targets

It is important to reiterate and emphasise the 'stock nature' of the climate change problem, i.e. that GHG concentrations, rather than emissions, matter. Global atmospheric CO_2 concentration has grown, rising from a pre-industrial value of about 280 ppm (parts per million) to 383.9 ppm in 2008, essentially by 150 years of industrialisation. According to the IPCC, they exceed by far the natural range of the last 65,000 years. There is an emerging political consensus that the stabilisation level should be somewhere between 450 ppm and 550 ppm. This would translate roughly into a 2°C or 3°C increase in temperature, which is a range that many consider as 'still manageable'. The EU set itself a target of 2° C. The level of the global target, whether expressed in terms of concentrations (i.e. 450 ppm) or temperature (e.g. 2°C), has implications for the projected emissions pathways, the scale of needed reductions and the time-frames. Due to the stock nature of the problem, early emissions reductions are beneficial both in environmental and economic terms. Or to put it differently, a delay in reductions would make it impossible to reach certain levels of stabilisation targets, and increase the marginal abatement costs as a result of a steeper future reduction curve. For example, a 3°C target requires the global peak some 15 years later, meaning that emissions need to peak far later for developing countries as well. For an overview, see Table 5.3.

Table 5.3 Overview of commitments by regions for different target scenarios

Scenario	Region	2020	2050
450 ppm	Annex I	-25% to -40%	-80% to -95%
CO_2 eq	Non-Annex I	Substantial deviation from baseline in Latin America, Middle East, East Asia and centrally-planned Asia	Substantial deviation from baseline in all regions
550 ppm	Annex I	-10% to -30%	-40% to -90%
CO_2 eq	Non-Annex I	Deviation from baseline in Latin America, Middle East and East Asia	Deviation from baseline in *most* regions, especially in Latin America, Middle East
650 ppm	Annex I	0% to -25%	-30% to -80%
CO_2 eq	Non-Annex I	Baseline	Deviation from baseline in Latin America, Middle East and East Asia

Source: Gupta et al. (2007, p. 776).

To date, governments participating in post-2012 negotiations are converging towards a long-term global target of a 50% emissions reduction by 2050 and towards 80% or more in 2100. For developed countries, halving global emissions by 2050 translates into an emissions cut in the range of 80-95% by 2050 and 25-40% in 2020 from the 1990 levels. By and large, even if developed countries make steep reductions (of up to 25-40% in 2020), it is without doubt that by 2020 or 2030 at the latest (depending on the target of 2°C or 3°C), emerging economies such as China, Brazil, South Africa and Mexico will have to reduce their emissions absolutely. It is indispensable that China will put in place effective GHG reduction measures in the future. By 2020, China's emissions are expected to be 30% higher than that of the US and by 2030, emissions from China will exceed the aggregate of OECD emissions. It is expected that a post-2012 regime will likely lead to further differentiation, e.g. developed countries (e.g. OECD), economies in transition (e.g. the former Soviet Union), emerging economies (e.g. China, Brazil, South Africa), most vulnerable countries (e.g. small islands) and least developed countries (LDCs). This means the extension of differentiation in allocation of responsibilities to the relations among developing countries.

More recent research goes a step further than the IPCC (2007). Table 5.3 shows that the 2°C limit target requires developing countries to limit their GHG emissions growth to 15-30% below projected baseline by 2020 (den Elzen & Höhne, 2008, p. 260; European Commission, 2009a, p. 5). The European Commission recommends in its Communication on the post-2012 regime (2009a, 2009b, 2009c, 2009d) that developing countries should aim at achieving the level through domestic measures alone.[42]

Developing countries are not expected to take on absolute targets under a post-2012 regime. However, in Bali in December 2007, developing countries accepted to take on "measurable, reportable, and verifiable" national mitigation commitments or actions on the condition that developed countries would provide assistance in the form of financing and technology. Hence, developing counties' commitments will remain confined to 'unilateral domestic actions' based on their national climate change or de-carbonisation strategy. This has been confirmed by the Copenhagen Accord.

5.3 Methodologies for allocation of responsibilities

The UNFCCC principle of 'common but differentiated responsibilities and respective capabilities' has been translated into works on developing methodologies to draw emissions pathways for different countries. This chapter looks at two specific proposals: the greenhouse development rights (GDRs) concept (see Baer et al., 2007; Kartha et al., 2008); and the Adaptation Financing Index (AFI) (see Oxfam, 2007).

---

[42] This excludes reductions triggered by crediting of emissions offsets but includes a rapid decrease in emissions from tropical deforestation. This also excludes emissions from international aviation and shipping, which account for around 4.5% (shipping) and 3.5% (aviation) of emissions from developed and fast-growing countries but represent the majority not under the Kyoto Protocol framework.

The GDR framework defines a 'development threshold'. Countries below this level are not expected to share the burden of mitigating the climate problem, as they have little responsibility for the problem and relatively little capacity to invest in solving it. Instead they are able to pursue their rightful priority of human development. With respect to those above the development threshold, the approach defines a burden-sharing framework based on the UNFCCC principle. As a result, the GDR index assigns obligations to contribute to the global mitigation effort – see the last column "Obligation (RCI)". The analysis then is applied to various countries including the US, EU and emerging economies. The GDR framework does not make assumptions on whether the obligation is fulfilled through domestic action or through reductions abroad, i.e. purchases of emissions rights. Table 5.4 shows a comparison of countries' or groups of countries' shares in various indicators.

For example, this framework calculates that the EU's obligation amounts to 27.4% of the global mitigation effort. Under a simple metric of current emissions, this figure would be at 11%.[43] To date, cumulative emissions of China for the period (1990-2005) amount to 13.8 giga-tonnes, which is not much different from those of the EU-27 at 17.8 giga-tonnes (and probably equal to those of the EU-15).

The same approach would bring about quite different results if the exercise is repeated in 2020, by which time China will most probably have been responsible for more cumulative emissions (since 1990 at least) than the EU, and probably also the US. On a 'business as usual' basis, China will probably have been the source of more cumulative emissions than the EU and US together (but probably still somewhat less on a per capita basis).

[43] In the year 2004, the global economy emitted about 49 billion tonnes of greenhouse gases (measured in CO_2-equivalent) (IPCC, 2007). The EU27's share was about 5.2 billion tonnes (UNFCCC, 2008) or 10.6%. Thus, *without* taking into account historical responsibilities related to past emissions, the EU would bear about 11% of global costs to combat climate change (see Behrens, 2008).

Table 5.4 Global shares of population, income, capacity, cumulative emissions, responsibility and obligation (RCI) for selected countries and groups of countries (%)

	Population	Income	Capacity	Cumulative emissions (1990-2005)	Responsibility	Obligation (RCI)
US	4.7	22.2	33.7	23.7	38.2	36.0
EU-27	7.6	23.2	30.0	17.8	23.5	27.4
UK	0.9	3.4	4.7	2.5	3.6	4.3
Germany	1.3	4.5	6.1	3.8	5.4	5.9
Russia	2.2	3.0	2.0	7.4	5.1	2.9
Brazil	2.9	2.8	2.2	1.3	1.1	1.7
China	20.4	10.0	2.3	13.8	3.4	2.7
India	17.0	4.2	0.1	3.8	0.1	0.1
South Africa	0.7	0.7	0.6	1.6	1.3	0.8
LDCs	11.6	1.5	0.1	0.4	0.0	0.0
All high income	15.6	59.1	83.4	52.7	79.4	82.3
All middle income	47.7	33.5	16.5	41.1	20.5	17.6
All low income	36.7	7.4	0.1	6.2	0.1	0.1
World	100%	100%	100%	100%	100%	100%

Source: Kartha et al. (2008).

Another formula is the Adaptation Financing Index (AFI) (Oxfam, 2007). This index focuses on adaptation. This index is based on four criteria (responsibility, equity, capability and simplicity) and respective indicators for the first three (excessive CO_2 emissions from 1992 to 2003, 2 tonnes of CO_2 per person per year[44] and the UNDP's Human Development Index (HDI)). It is important to note that the index covers only countries that are both responsible and capable and gives equal weight to responsibility and capability (50% each). For example, EU responsibility amounts to 32%, the

[44] This takes into account the target of halving 1990 CO_2 emissions levels (10.7bn tonnes) by 2050 and the average size of the global population from 1992 to 2003 (Oxfam, 2007).

US 44% and Japan 13%. While China has average per capita emissions of 2.7 tonnes from 1992 to 2003, China's HDI is low and therefore considered not to have the capacity to assist in financing for adaptation.

Such indices are useful analytical tools to establish some numerical indicators for allocation of responsibilities to ultimately construct possible emissions pathways. However, they are not sufficiently tested or politically accepted to serve as the basis for target-setting in UN negotiations.

5.4 Methodologies and data for comparability of efforts

The principle of 'common but differentiated responsibilities and respective capabilities' provides a guideline for top-down allocation of responsibilities, but not more. Efforts to operationalise the principle typically collide with countries claiming the right to receive special treatment because of national and specific circumstances.

It is fair to say that the discussion on 'comparability of efforts' with implication for 'equivalence of measures' among developed counties is more advanced and methodologically easier than the one among developing countries. A number of indicators could be used to narrow down the margin of error of what ultimately will be a political decision based on negotiations. Examples of such indicators include: i) per capita emissions; ii) GHG intensity (GHG emissions per unit of GDP) adjusted by climate, geography, industrial structure, population, availability of fuel or marginal abatement cost calculations; and iii) per capita GDP. The European Commission (2009a) proposed four indicators: per capita GDP, GHG intensity, trend in GHG emissions from 1990 to 2005 and population trends from 1990 to 2005.

Developing countries fear that common metrics for comparability may eventually lead to a back door for target-setting, not foreseen neither by the UNFCCC nor the Kyoto Protocol. Generally, therefore, they are more relaxed with bottom-down approaches that would recognise and possibly reward voluntary initiatives of individual countries (see, for example, McKinsey & Company, 2009). On the other hand, there is a strong economic argument for major contributions of developing counties. Typically, marginal abatement costs in developing counties are significantly lower than in developed countries. Focusing too much on higher-cost developed countries will make total global compliance costs higher and therefore reduce global welfare.

A number of developing countries and some emerging economies, including BRICs, undertake considerable efforts to reduce GHG emissions. In many cases, such initiatives relate more to local pollution and energy security, or access to energy. These actions impose costs on industry and consumers. Hence, they create a shadow carbon price. Given that developing countries are not ready to agree on common indicators, such a shadow carbon price would be the second-best tool to compare 'equivalence' in efforts of individual countries in a systematic way.

Efforts of emerging economies such as China, Brazil, Korea, Mexico, South Africa or China are considerable and costly. For example, China and India regularly claim that they spend up to 2% of GDP or even more per annum on climate change. While this includes adaptation measures, a big part nevertheless goes into mitigation. Shying away from taking economy-wide GHG reduction commitments – especially if legally binding under internal law – they undertake a wide range of domestic measures. Typically they include energy-intensity reductions, renewable energy targets or policies to insulate or retrofit the housing stock. Such measures are seen as being broadly beneficial to energy-security concerns. However, emerging economies tend to avoid setting GHG reduction targets in order not to jeopardise the use of coal for power generation and to avoid having to phase out politically delicate fossil fuel subsidies or tackle the transport sector, all of which are seen as vital for economic growth and social stability.

China is but one of the emerging economies that is taking substantial measures. Its position on climate change has substantially changed recently realising that its current high-carbon growth model is not sustainable much beyond 2020 or 2030. Increasingly policy-makers are becoming aware of the links between climate change, energy security and air pollution policies. This materialised in China's pledge in the context of the Copenhagen Accord to take voluntary autonomous domestic mitigation[45] action. Thus,

---

[45] All developing country pledges explicitly refer to Article 4.7 of the UNFCCC, which states that nationally appropriate mitigation actions are voluntary and their implementation depends on financial, technological and capacity-building support by developed countries.

according to the letter [46] to the UNFCCC Secretariat, China will "endeavour" to reduce its CO_2 intensity (CO_2 emissions/unit of GDP) by 40-45% compared to 2005 by 2020. The pledge also includes increasing the share of non-fossil fuels in primary energy consumption to around 15%, forest coverage by 40 million hectares and forest stock volume by 1.3 billion cubic meters over the same time horizon.

These actions are a continuation of the domestic policies and measures already in place (e.g. energy efficiency targets in the 11th Five Year Plan and the 2007 National Climate Change Programme) for 2006-10, including a target to reduce energy consumption per unit GDP by 20%, to raise the proportion of renewable energy (including large-scale hydropower) in total primary energy consumption up to 10% or to increase the forest coverage rate to 20% and to increase carbon sinks by 50 Mt over the level of 2005. According to the China Sustainable Energy Programme (2008), these measures will amount to 1500 Mt CO_2-eq emissions avoided over five years. The World Resources Institute (WRI) in Washington, D.C. reports that China may be able to reach the target, as energy intensity has decreased by 1.8% in 2006, by 3.7% in 2007 and by 4.2% in 2008, and is thus close to the 4% annual goal (Seligsohn & McMahon, 2009). Some policies are comparable to developed countries' actions. For example, the Renewable Energy Law from 2005 sets a legally binding 15% share of renewable sources in primary energy by 2020 and is supported by many incentives, including a form of 'feed-in' tariffs for wind (China Sustainable Energy Program, 2008). Wind power receives support of 0.7 yuan per kWh (about 0.5 US cents) (Yan & Jiang, 2009). More specific and operational objectives are stipulated within the various regulatory actions, in areas such as improving fuel economy in transport, energy diversification, reforestation and energy-efficiency enhancements in both power generation and end-use (buildings and household appliances). China's National Energy Plan also sets a target of total nuclear power generation at 40 GW by 2020 (NDRC, 2005).

---

[46] This and the other letters with pledges by China, India, Brazil, Mexico, Korea and South Africa can be found under "Appendix II - Nationally appropriate mitigation actions of developing country Parties" on the UNFCCC's website (http://unfccc.int).

Similarly, the South African government has pledged, in the context of the Copenhagen Accord, nationally appropriate mitigation actions that would 'enable' a 34% reduction in emissions growth relative to a 'business as usual' trajectory by 2020, and a 42% reduction by 2025. It also confirmed the so-called 'plateau and decline' trajectory adopted in 2008 that could lead to around -30 to -40% by 2050 compared to 2003.

In the context of the Copenhagen Accord, Mexico stated its aim to reduce GHG emissions by up to 30% compared to business-as-usual by 2020, conditional on support by developed countries, and reiterated that its 2009 Special Climate Change Program is expected to achieve GHG reductions of 51 $MtCO_2eq$ by 2012.

The situation is different for India, with its low per-capita GHG emissions of 1.7 tonne in 2005 (1.1 tonne for CO_2 only). Nevertheless, in the context of the Copenhagen Accord, India has made a pledge to reduce the emissions intensity of its GDP by 20-25% by 2020 from 2005 levels (excluding agricultural emissions) through domestic actions of a non-binding character. This builds on India's past achievements to partially de-couple economic growth from the growth in GHG emissions and to keep pace with annual growth in energy demand at almost half the pace of economic growth[47] in recent years. Further actions are envisaged, such as an objective of installing solar-power generation capacity of 20,000 MW by 2022, mandatory renewable purchase obligations (a form of feed-in tariffs) and energy efficiency initiatives that are expected to reduce energy consumption by 5% by 2015 (Fujiwara & Egenhofer, 2010). In addition to mandatory energy efficiency standards and labelling, these initiatives include a domestic trading scheme of energy-efficiency certificates that covers large energy-intensive industries. This trading may to an extent develop into a shadow carbon price.

This shadow carbon price is difficult to assess and in fact mainly consists of many different shadow prices in the various sectors. Typically, developing countries undertake sectoral policies starting with those sectors in which emissions can be monitored, policy instruments work best, and

[47] Own calculations for the period 2000-07 that show 7.3% annual GDP growth on average and 3.8% annual primary energy demand growth on average, based on data for from the IEA (2009b) and the IMF (2009).

there are benefits in terms of emissions reductions, energy security or reduction of local pollution.

Consequently, assessing whether there is at least an implicit carbon price in developing countries is difficult and can be very tentative at best. One possible option for carbon pricing would be to assess and compare the costs of countries' actions, for example, against standard costs curves such as the ones from McKinsey & Company and the IPCC. But that assumes that data are freely available. Based on such comparisons, one could make some very rough estimates on the costs of the measures. Divided by the estimated reductions of a measure, one could then estimate the cost per tonne of avoided CO_2. Doing this in a bottom-up fashion for all the policies could give some indication of an average CO_2 cost, or the shadow carbon price. While this methodology is necessarily very crude and may not pass WTO scrutiny, it nevertheless could serve as an analytical tool. A complication is that all emerging economies face big deficiencies in the collection of data as well as in measuring, reporting and verification (MRV) of emissions, which would require some capacity-building.

5.5 Concluding remarks

The UNFCCC lays down the principle of 'common but differentiated responsibilities and respective capabilities'. This means that developed countries have more responsibility than developing countries and should take a lead in climate action. To date developing countries argue that the differentiation part of the principle only applies to the relations between developed and developing countries – not those among themselves. This makes it almost impossible to develop a global cap-and-trade system encompassing all major economies. Some attempts have been made to develop specific methodologies with key indicators for quantification and allocation of responsibility, e.g. the Green Development Rights (GDR) and the Adaptation Financing Index (AFI).

On the other hand, this principle does not prevent countries from taking their own actions, and the UNFCCC even states such action as part of a general commitment by all countries. Hence, developing countries prefer unilateral – internationally non-binding action based on domestic policies and measures reflecting their own national circumstances and priorities. This explains to some extent the current large difference in the price of carbon. As long as these countries are not keen or unable to discuss common indicators for comparability of efforts with developed countries or among themselves, the second-best tool could be found in a shadow carbon price.

There are numerous climate change mitigation efforts by developed countries and a number of developing or emerging economies, such as China, Russia, India, South Africa and Mexico, which may lead to significant carbon costs and may possibly even provide a (shadow) carbon price. Although it is preferable to translate these actions into estimates of total abatement and associated costs, this would be possible only in exceptional cases and on a very tentative basis, given the constraints on data collection and MRV. To give a more accurate account would require better data, thorough MRV, and a more in-depth analysis of national strategies.

6. ENFORCEMENT AND COMPLIANCE

As Chapter 5 illustrates, both the UN Framework Convention on Climate Change and the Kyoto Protocol distinguish between two groups of countries:

i) developed countries and economies in transition (=Annex I countries) and

ii) developing countries.

Under the UNFCCC, these two groups have common responsibilities, such as submission of national communications, but they also have different responsibilities. Only the former (Annex I countries) have taken on firm quantitative emissions reduction commitments under the Kyoto Protocol. This is likely to change for the post-2012 period. While Annex I countries are expected to accept bigger absolute GHG emissions reductions, emerging economies are supposed to undertake national mitigation actions to reduce emissions below business as usual. Developing countries' actions would have to be binding and 'measurable, reportable and verifiable' (MRV) while supported by finance and technology by Annex I countries.

This chapter discusses the compliance and enforcement mechanisms of the Kyoto Protocol for commitments by Annex I countries. Currently, there are no enforcement mechanisms for pledged domestic actions by non-Annex I (i.e. developing) countries, and the commitments to robust MRV systems are so far also in a state of flux. It will thus first look at major Annex I countries' Kyoto Protocol commitments and their progress in emissions pathways and then examine how the UN compliance mechanism or enforcement tools operate.

6.1 The Kyoto Protocol in the first commitment period (2008-12)

The Kyoto Protocol is an international agreement to substantiate the UN Framework Convention on Climate Change (the Convention). The major feature of this Protocol is to set binding targets for greenhouse gas emissions from 37 developed countries and the European Community. While the Convention encourages developed countries to stabilise GHG emissions, the Protocol commits them to do so. The Protocol places a heavier burden on developed countries in terms of quantitative emissions reduction targets, but it also requires all countries to meet general commitments.

6.1.1 *Commitments, achievements and prospects for compliance*

Under the Kyoto Protocol, Annex I countries are committed to undertake reductions of six greenhouse gases: carbon dioxide (CO_2), methane (CH_4), nitrous oxide (N_2O), hydrofluorocarbons (HFCs), perfluorocarbons (PFCs), and sulphur hexafluoride (SF_6). The first commitment period lasts from 2008 to 2012. Each party's compliance with the targets will be assessed compared to GHG emissions in 1990 or, in case of 'economies in transition' (EITs), in the base year or period multiplied by five. Some countries can include in their 1990 emissions base year or period the aggregate GHG emissions by sources minus removals by sinks in 1990 from land-use change, i.e. taking LULUCF into account by calculating their assigned amount.

Annex I countries, mainly developed countries and EITs, committed themselves to collective emissions reductions of 5.2% from 1990 levels in the period 2008-12. These cuts were divided within the group, for example, 8% for the EU-15, 7% for the US and 6% for Japan.

Table 6.1 provides the key data, which highlight a significant variance in Annex I countries' commitments and achievements over the 2008-12 period. It shows that among the Annex I countries, Russia and Ukraine are outliers, both in terms of their low level of GDP per capita and their huge reductions in absolute emissions (due to the collapse of Soviet heavy industry).

Table 6.1 Key statistics of selected Annex I parties

	GDP/ capita 2007[a]	Ratified the Kyoto Protocol	Total GHG emissions 2007, excl. LULUCF and bunker fuels ($MtCO_2e$)	Change in emissions 1990 -2007	Kyoto target (1990[e] - 2008-12)	Distance from target 2007
EU-27[c]	34,118[b]	31/05/02	5,045.1	-9%	-	
EU-15[d]	40,066[b]	31/05/02	4,052.0	-4%	-8%	-4%
Australia	42,864	12/12/07	541.2	30%	8%	-22%
Canada	43,404	17/12/02	747.0	26%	-6%	-32%
Japan	34,287	04/06/02	1,374.3	8%	-6%	-14%
Russia	9,103	18/11/04	2192.8	-34%	0%	34%
Turkey	9,422	-	372.6	119%	-	-
Ukraine	3,100	12/04/04	436.0	-53%	0%	53%
US[f]	46,674	-	7,107.2	17%	-7%	-24%

[a] Data for 2007, US$ per capita, current market prices, from IMF (2009).

[b] Population data from Eurostat; GDP data from IMF (2009) for 2007, US$, current prices.

[c] EU-27 emissions data from EEA (2009, p. 9), includes Malta and Cyprus, which have no targets under the Kyoto Protocol.

[d] The 15 states that were EU members in 1990 will redistribute their targets among themselves, taking advantage of a scheme under the Protocol known as a 'bubble', whereby countries have different individual targets, but which combined make an overall target for that group of countries. The EU has already reached agreement on how its targets will be redistributed.

[e] Some EITs have a baseline other than 1990.

[f] The US has indicated its intention not to ratify the Kyoto Protocol.

Note: Although they are listed in the Convention's Annex I, Belarus and Turkey are not included in the Protocol's Annex B as they were not Parties to the Convention when the Protocol was adopted.

Source: Based on data from UNFCCC website and database (http://unfccc.int/), EEA (2009), IMF (2009) and Eurostat.

Among those on the list overshooting the targets (on the basis of 2007 data), only the US did not ratify the Protocol. All others, including Canada, Australia and Japan, did. However, those likely to overshoot are with only one exception committed politically to make up for the excess emissions by acquiring offsets, either in the form of credits under the CDM (Clean Development Mechanism) (see Chapter 1, Introduction) or by acquiring credits from countries with so-called 'hot air', i.e. essentially the group called 'economies in transition'.

Russia and Ukraine are the two countries with the largest amounts of excess credits (see Table 6.2). However, they have not been able to sell large amounts of their 'hot air'. Several reasons are thought to be behind this. Delays in the required domestic infrastructure have played a role, as has the reluctance of the Annex I countries most in need of credits or offsets to satisfy their Kyoto commitments this way. It has also been suggested that Russia does not want to sell massive amounts of credits, preferring to carry forward its existing credits to future periods.

A number of potential buyer countries in Annex I are known to have set up budget lines to finance these credits some time ago, but little is known about the extent of both the purchases already undertaken and those planned for the immediate future. However, there seems to be no reason to doubt the political will of most countries to honour their Kyoto commitment.

The exception to this rule is Canada where successive governments have not taken any steps to ensure that the country can meet its commitment under Kyoto. Moreover, there seems to be a strong domestic consensus that the country is not willing to acquire the credits needed to offset the likely excess emissions of the country. Canada thus constitutes the only test case for the compliance mechanism under Kyoto. The key question now is what will be the consequences of non-compliance with the Kyoto Protocol.

Formally Canada committed to reduce GHG emissions to 6% below 1990 levels in the period 2008-12, equivalent to an annual 'allowance' of 548.4 mega (=millions) tonnes of carbon dioxide equivalent (Mt CO_2eq). In 2007 Canada's GHG emissions amounted to 747 Mt CO_2eq, which is about 26% over its 1990 emissions levels and almost 200 Mt CO_2 equivalent (33.8%) above its Kyoto target (Environment Canada, 2009). The overall trend until 2007 was one of steady growth of emissions instead of the required decline (see Figure 6.1).

Figure 6.1 Canada's GHG emissions 1990-2007

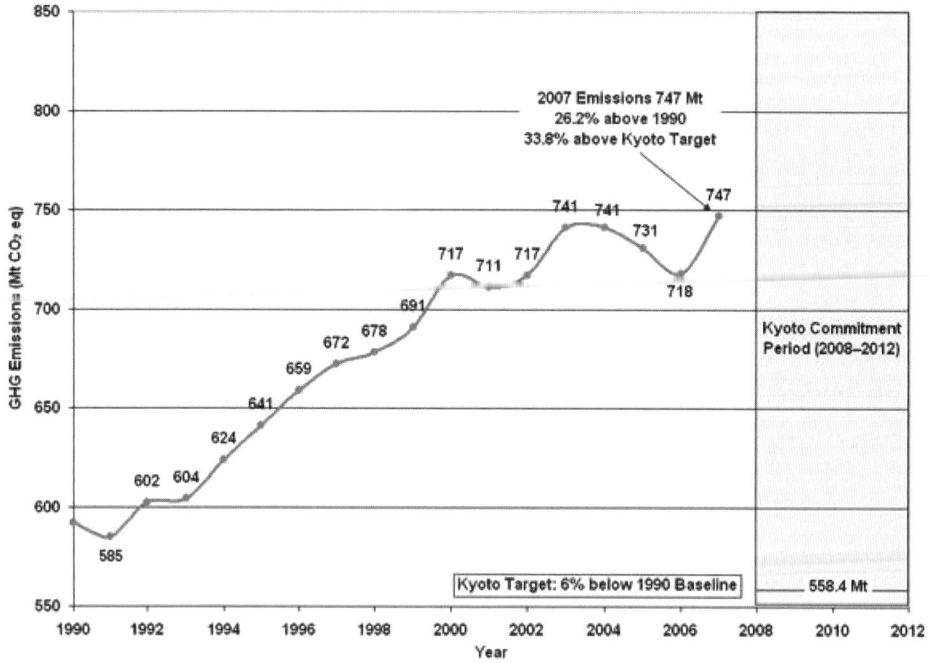

Note: Recent fluctuations have been attributed to the demand for heating fuel during winter seasons, which has grown warmer except for the 2007 winter, which was much colder (Environment Canada, 2009).

Source: Environment Canada (2009).

For the moment the Canadian government emphasises that the country remains committed to its reporting requirements under the UNFCCC (and the Kyoto Protocol). However, given that mitigation policies have only been put in place in recent years, its focus has been shifted from the remaining years of the first commitment period (2008-2012) to the 'future': the goal of a 20% GHG emissions reductions from the 2006 level by 2020 and a 60% to 70% reduction from the 2006 level by 2050.[48] However, such a shift in the baseline would not be covered under the Kyoto Protocol. This raises the question of how the KP compliance system will work in reality.

---

[48] For more details, see the website of Environment Canada, a department of the Canadian government responsible for coordinating environmental policies and programmes (http://www.ec.gc.ca).

6.1.2 *How does the Kyoto Protocol compliance system work?*

The Kyoto Protocol contains a compliance mechanism (Article 18) in order to ensure that parties meet their commitments. Specific procedures and mechanisms were introduced in the first meeting after the entry into force of the Protocol (Decision 27/CMP.1). The Compliance Committee has been operational under this framework since March 2006. This committee is composed of two branches: the facilitative branch and the enforcement branch. The former provides technical and financial advice to parties facing difficulties in meeting their commitments. The latter ensures that the parties comply with their commitments. Compliance cases that can be currently brought before the enforcement branch are limited to matters of procedure or implementation (e.g. registry, reporting). To date there have been three cases – brought against Greece, Canada and Croatia as regards compliance in national inventory reporting, national registry and assigned amount. This has led to declarations of 'non-compliance', but the countries remedied the situation the cases were subsequently closed.

A case of non-compliance with emissions targets cannot come before the enforcement branch until after the end of the commitment period in 2012.

Formally three major consequences for non-compliance are foreseen in the KP:

1) A country in non-compliance with the 2012 target has 100 days after the expert review of its final emissions inventory to make up any shortfall (i.e. to buy credits).
2) If the country still misses its target, it must make up the difference, plus 30% in the second commitment period after 2012.
3) The country will also be suspended from making transfers under emissions trading (i.e. the transfer of AAUs, CERs and ERUs) and within three months it must submit a plan on the action it will take to meet targets for the second commitment period.

How would these mechanisms work in the case of Canada? Canada would not be much affected by being suspended from making transfers under emissions trading. However, the first two compliance mechanisms would have substantial consequences (if adhered to):

1) *Making up the shortfall by buying credits.* How many credits would Canada have to buy? An order of magnitude can be calculated assuming emissions stabilise at the 2007 level, which is roughly 200

MT above the target. In this case Canada would have to buy roughly 1,000 MT (5 times 2,000 for the five years 2008-12).[49] The cost of this would of course depend on the price of emissions in 2013, which is difficult to predict today. Moreover, the question that will then arise is from whom would Canada buy these credits? Wherever they are cheapest? Via CDM credits? A good guess is that by 2013 (i.e. in the new commitment period) the price of emissions permits should be around $30, which seems to be the upper acceptable limit in the US. The EU price might be somewhat higher, but Canada would certainly prefer the cheaper US source. This would imply a cost for Canada of $30 billion (about 2% of GDP). It is highly unlikely that the Canadian government would agree to pay such an amount.

2) *Making up the difference plus 30% in the second commitment period.* This seems also difficult to implement: Canada would need to make up 1,300 MT or 260 MT on a per annum basis. Taken literally this would mean that over the second commitment period Canadian emissions would have to fall by over 30% just to satisfy the country's obligation from the first period.

It thus remains to be seen how the Canadian case will be dealt with. The Protocol states that any procedures and mechanisms for compliance entailing binding consequences shall be adopted by means of an amendment (Article 18). Agreement on such an amendment has not been reached due to strong opposition, thereby weakening the enforcement function. This leaves the current compliance mechanism mainly operating in a facilitation mode.

If, as is likely, Canada does not face any substantial financial or other penalty, the compliance mechanism of the Kyoto Protocol will have become a paper tiger.

In summary the Kyoto Protocol currently falls short of acting as an effective and robust compliance mechanism in these respects:

---

[49] This simple calculation also shows that the concentration on the 'commitment period' does not take into account the 'excess' emissions in the previous years. This does not make sense given the stock nature of the problem (CO_2 concentrations matter, not emissions).

1) No preventive action can be taken against parties that will most likely fail to comply in terms of emissions reductions before the end of the current commitment period.
2) There is no financial penalty or any consequence leading to the loss of credits, and the impact of suspension for eligibility in participating in flexible mechanisms will be unclear until emissions targets are set and available flexible mechanisms are identified for the second commitment period.
3) Whatever cases are brought before the enforcement branch of the Compliance Committee, the decision will not entail binding consequences until the Protocol is amended.

6.2 Post-2012 discussions

It is somewhat surprising that reform of the enforcement mechanism has not been a major issue during the preparations for the post-2012 regime. Instead the diplomatic action has concentrated on numerical targets and the size of financial transfers. Major developed countries have declared their short- and mid-term emissions reduction targets (see also Chapter 5), while developing countries insist that the Convention and the Kyoto Protocol allow them to stay clear from target-setting.

Nevertheless, the term 'binding commitment' is widely used in the preparations for the post-2012 regime. But in reality it seems to indicate a political commitment, rather than something that could be enforced via a legal mechanism. As for the compliance mechanism, the relevant subsection in the negotiating text appears to be largely based on old Article 18, Kyoto Protocol (see 1.1.2). At the time of writing, it is difficult to discuss the nature of the self-declared emissions reduction targets unless they are made legally binding under domestic legislation, as in the case of the EU.

6.2.1 *Ensuring enforcement of pledged domestic actions by developing countries*

Most non-Annex I countries are unlikely to take on binding emissions targets and will seek to maintain the existing flexible mechanisms such as the Clean Development Mechanism. However, advanced developing countries are finding themselves under pressure from developed countries and have become more open to pledging domestic actions and participating in international mechanisms that would reward their early actions. Given little chance for application of target-and-timetable approaches under the Kyoto Protocol to developing countries, it may be worthwhile to expand the scope of our enquiry to a new flexible mechanism that could ensure enforcement of pledged domestic actions by more advanced developing countries.

As envisaged under the 2007 Bali Action Plan (paragraph 1(b) (ii)), developing countries' actions are called Nationally Appropriate Mitigation Actions (NAMAs). Taking the process one step further, some attempts have been made to develop new flexible mechanisms based on NAMAs and to reward early actions taking place beyond project levels, either on the economy-wide or sector basis (from key sectors of economy). A number of options involving sectoral trading, sectoral crediting and NAMAs crediting or trading, either separately or in combination, have been under discussion on the Convention track (AWG-LCA).

If one of these options materialises in the absence of a comprehensive compliance system, an enforcement tool might be attached to this specific market mechanism. However, discussions on the Convention track (AWG-LCA) suggest that in all likelihood there will be no binding consequences of developing countries' under-performance. Thus it is likely that such a new flexible mechanism would be set up without detailing how enforcement tools work.

6.3 UNFCCC and the Kyoto Protocol on trade measures

The UNFCCC has a single stand-alone provision in Article 3.5 on trade measures,[50] which was inspired by the chapeau of Article XX, GATT (for Article XX, see Chapter 4.2.2). The UNFCCC does not refer or defer to the WTO as the mechanism with the authority to interpret this provision or to assess the legality of trade measures placed by parties (Werksman & Houser, 2008). Parties have not resorted to using the dispute settlement process envisaged in Article 14 of the UNFCCC. In contrast to the

---

[50] "The Parties should cooperate to promote a supportive and open international economic system that would lead to sustainable economic growth and development in all Parties, particularly developing country Parties, thus enabling them better to address the problems of climate change. Measures taken to combat climate change, including unilateral ones, should not constitute a means of arbitrary or unjustifiable discrimination or a disguised restriction on international trade."

UNFCCC, the original chapeau of GATT Article XX is supported by a number of specific provisions as well as the Panel procedures.

Moreover, it appears that Article 3.5 neither endorses nor prohibits using trade measures equivalent to those in the Montreal Protocol or other agreements as a means of increasing the effectiveness of the Convention in terms of compliance and enforcement (Yamin, 2004). For example, the Montreal Protocol on ozone depletion identifies a number of trade measures: i) requiring parties to ban the trade of controlled substances and to implement a licensing system and measures to control the trade of products using these substances, and ii) banning the import of HCFCs from non-parties from 2004 and the trade in bromochloromethane with non-parties as of 2001. Furthermore, trade measures have been introduced in selected multilateral environmental agreements (MEAs) including the Basel Convention on hazardous wastes, the Rotterdam PIC (Prior Informed Consent) Convention as well as the Montreal Protocol.

Article 3.5 is one of the five guiding principles of the UNFCCC in the same status as the other principle on common but differentiated responsibilities and respective capabilities, Article 3.1. Nonetheless, it appears that the key principle on use of trade measures has been taken out from the original GATT context and subject to re-interpretation in the UNFCCC mind-set as the Indian proposal shows below.

Under the UNFCCC and the Kyoto Protocol, developed countries commit themselves to minimising adverse economic, social and environmental impacts on developing countries when responding to climate change (Article 4.8, the UNFCCC; Arts. 2.3 and 3.14, the Kyoto Protocol). In this respect the UNFCCC does not mention trade measures (4.8) whereas the Protocol makes an explicit reference to effects on international trade as one type of adverse effects to be minimised (2.3). In the course of subsequent decisions and discussions, trade measures have been addressed in a broader context of Annex I countries' implementation in general, and consequences of their 'response measures' in particular. Consequences of developed countries' actions on developing countries matter. Not the other way around.

The Indian proposal for prohibiting unilateral measures has received a wider support from emerging and other developing economies, including China, Brazil, Argentina, Venezuela, Singapore, and Saudi Arabia, and has met opposition from developed countries such as the US and Canada (IISD, 2009a, 2009b). One possible breakthrough for easing this tension, however, might be found upon return to clarification about interpretation of Article 3.5, complemented by additional legal provisions and enhanced institutional arrangements. Since a dispute settlement process is not functioning in the UNFCCC, the Convention parties that are also parties to the WTO could take the case to the WTO panel. The WTO would then likely look to the UNFCCC for guidance on an appropriate standard for the comparability of efforts to reduce emissions as well as for an appropriate standard for an assessment of the compatibility of the trade measures with the GATT Article XX and UNFCCC Article 3.5 (Werksman & Houser, 2008).

From the UNFCCC perspective, there will be two challenges. One is how to agree on any interpretation of Article 3.5, given that the provision has been linked to a broader debate on adverse effects, as has been discussed. Another is who could provide such a guidance.

6.4 Summary and concluding remarks

Developed countries (Annex I Parties) have produced a mixed record on commitments and achievements, almost halfway into the 2008-12 commitment period. Since compliance is assessed over the whole of this five-year period, it is premature to determine the status of parties' compliance at this stage in a formal sense, and the economic crisis has added to the uncertainty of the outcome in terms of actual emissions. The available evidence suggests that in 2008 and 2009, emissions fell considerably, at least relative to 2007. Given that the recovery is likely to be slow, this implies that the crisis should have made it much easier for most developed countries to honour their commitments, especially given that many of them had already purchased substantial credits before the crisis started. For more details, see also the next chapter.

There seems to be one exception to this, namely the case of Canada, whose recorded GHG emissions were in 2007 26% over 1990 levels and thus 33.8% over the Kyoto Protocol target. Canada's behaviour will thus constitute a key test case for the compliance and enforcement mechanism of Kyoto. It seems at present that the country will not face substantial consequences for its non-compliance. This illustrates a key weakness of the Kyoto Protocol, which adopted a target-and-timetable approach without a robust compliance system or effective enforcement tools.

The Kyoto Protocol's overall compliance system is unlikely to be amended since this would require agreement by all parties. Both the

Framework Convention and the Protocol foresee the creation of new flexible mechanisms such as NAMAs crediting or/and trading. NAMAs were devised as a concept to acknowledge pledged actions by developing countries. Enforcement tools for NAMAs crediting and trading would be facilitative measures that apply to developing countries' under-performance, but there will be no legally binding consequences of their under-performance. The decision over the modalities and procedures relating to NAMAs will take place only after a Copenhagen agreement is reached. The expected continuous lack of the compliance system or enforcement tools suggests that there will be no institutional framework to ensure enforcement of developing countries' mitigation actions, except through a 'carrot' of assistance in finance and technology.

The UNFCCC has a provision on trade measures based on the GATT Article XX. On the other hand, the discussion on trade measures has been framed in a wider context of adverse effects of developed countries' response measures on developing countries, which resulted in a tension between the two sides. One breakthrough could be found upon return to clarification about interpretation of Article 3.5, to be strengthened by additional legal provisions and enhanced institutional arrangements.

Looking ahead into the post-2012 period raises a fundamental problem for a partner, like the EU, that has undertaken substantial commitments and whose industry is paying a substantial price in terms of higher costs in order to honour the commitments undertaken.

Sovereign states are always reluctant to submit themselves to binding international enforcement mechanisms. The weakness of these mechanisms in the Kyoto Protocol is thus not a surprise. The approach that has been followed by the EU (and almost certainly by the US as well) is to make their own commitments binding domestically. In the case of the EU, this has been done via EU directives which bind all member countries. Similarly, the US seems intent on making its own targets legally binding within its own constitutional framework - but also to avoid any legally binding international commitment.

But what can countries that have taken their own binding commitments do to ensure that the rest of the world does not free-ride on their efforts? Border measures seem to remain the only effective enforcement mechanisms, given that both the EU and the US are the world's two largest markets.

7. CONCLUSIONS

This study has focused on a concrete question, namely whether it is appropriate to use 'border measures' to complement a domestic cap-and-trade system like the EU emissions trading system, complemented by some form of protection against carbon leakage. The point of view taken in this study has not been that of defending a level playing field for particular sectors. In our view the key issue is whether a carbon import tax is desirable from the point of view of global welfare. The main finding is that this is indeed the case.

This finding in turn raises two practical questions: how can the EU take a decision to impose such a tariff and how high might it be?

The response to the first question is straightforward: The EU exercises exclusive competence for all matters concerning the customs union. Any decision to impose a carbon border tax would have to start with an initiative by the Commission which would then need to be approved by the Council and the European Parliament. Approval in the Council would require only a qualified majority.

The second issue, the size of the import tariff, is more complicated. The key question here is what rate of a CO_2 import tariff would be appropriate if the aim is to maximise global welfare? There is surprisingly little literature on this question, as most contributions to this debate have focused on the impact of carbon border taxes on the competitiveness of specific industries (mostly those subject to cap-and-trade systems). The implicit assumption in most of the literature seems to be (see in particular Houser et al., 2008) that the tariff rate might be high for the exposed sectors (cement, steel, aluminium, chemicals, pulp and paper) but rather low for other sectors. As trade in the energy-intensive sectors amounts to less than 5% of all trade (for the US, but also for the EU), it has often been argued that competitiveness concerns based on the distorting impact of domestic carbon prices without border measures are exaggerated. However, most observers have concentrated on the few energy-intensive sectors covered by cap-and-trade systems, which has had the effect of obscuring the general equilibrium effect of a carbon price. The latter would affect all sectors in which energy or fossil fuels are used as an input.

Some rough, preliminary calculations suggest immediately that a carbon tariff by the EU (or the US) could be much higher than the average most-favoured nation tariff rates of the EU (on average about 3-4%). Recent calculations by Weber et al. (2008) suggest that the total CO_2 embodied in China's 2005 exports (in jargon: EEE, or 'embodied emissions in exports') should be around 1,670 million tonnes of CO_2, or over 30% of all Chinese emissions. This percentage corresponds roughly to the share of exports in the Chinese economy (around 35%). Given total Chinese exports in 2005 of around $760 billion, this implies an *average* carbon intensity of a little more than two tonnes of CO_2 per $1,000 of exports from China.

The next question is the level of the carbon price. This is very difficult to estimate. It is tempting to use the price under the present ETS. However, this price refers only to the present Kyoto regime and cannot serve as a guide to what would result under the post-Kyoto regime. In the post-2012 period, it is likely that the constraint will be much tighter. The Kyoto commitment (of an 8% reduction compared to 1990 levels of emissions) did not represent a tight constraint for the EU (15), given that the collapse of industry in the former East Germany on its own caused a reduction of about 3-4% in emissions and the fuel switch from coal to gas as a result of electricity and gas market liberalisation also led to reductions in the UK. Moreover, the current recession, which is likely to affect at least three years (2008-10) of the five-year commitment period, is reducing energy use and thus emissions even further, relative to a baseline of 2% potential growth.[51] Some studies (for example, Amann et al., 2009; Spencer et al., 2010) suggest that the EU 2020 GHG reduction targets of 20% will be met by the emissions reductions as a result of the economic crisis. Consequently, the studies find – by and large – that a post-crisis 30% reduction target would

---

[51] Emissions in the sectors covered by the EU ETS have dropped by 11% in a single year (2008-09) (for more information, see the European Commission's webpage on "Registries and Community Independent Transaction Log" at http://ec.europa.eu/environment/climat/emission/citl_en_phase_ii.htm).

be required to match the assumed constraint for a 20% reduction prior to the crisis.

Assuming that the EU moves to a unilateral 30% reduction target, the EU constraint would naturally become tighter again. Pre-crisis, the Commission had estimated that a carbon price of around the €40-50 per tonne would have been required to reach the EU's 2020 commitments of 20% reduction. Given that the crisis might have reduced the 2020 level of EU GDP by about 10% (at least compared to the pre-crisis expectations), one could argue that a post crisis target of -30% should be equivalent to the pre-crisis target of -20%. One would thus expect that the above carbon price would be reached only if the EU moved to a 30% reduction target now. This would imply, at current exchange rates, about $50-60 per tonne. This is most likely too high for the US, where $30-40 per tonne has been estimated to constitute the upper limit of what is politically acceptable. Thus, taking $40 as a rough guide (or, roughly €25-30 per tonne), this means that a border carbon tax on Chinese exports (to the EU) would be a bit more than two times $40 per $1,000 of exports, or approximately 8-9%. As China upgrades the sophistication of its exports, this rate might come down, but under current conditions the average carbon tax could thus be very significant, much higher than the MFA tariffs currently applied by the EU, and certainly an order of magnitude larger than the modest tariff reductions that were contemplated under the Doha round.

A comparison of carbon intensities across other countries shows immediately that in general the BRIC countries, which represent the major emerging economies, have a much higher carbon intensity of exports (ranging from about 1 tonne of CO_2 per $1,000 of exports for Brazil to about 4 for Russia) than OECD countries (around 0.5). This implies that the average carbon import taxes the EU might have to impose on imports from these countries would thus range from about 4% for Brazil to 16% and more for Russia, but only around 2% for any OECD countries without comparable domestic measures.

It is apparent that the imposition of such relatively high tariffs would have important implications for trade flows. This is actually desirable as some trade flows incorporate substantial carbon. Nevertheless, it is clear that a first-best solution would involve action by major advanced developing nations. Tirole (2009a and b) also emphasises this point. So far, however, developing countries have explicitly ruled out making any binding commitments on caps. But there is reason to believe that the mere consideration of a carbon import tax by major importers could make the first-best solution actually possible. It is sufficient to consider the incentives for developing countries if the EU (jointly with the US) were to impose (or just consider imposing) a carbon border tax. At that point, developing countries would have the choice between taxing carbon at home or having it taxed by the EU at its border. The former would of course be preferable from their perspective because they could then collect the revenues. Just considering the imposition of a carbon border tax could thus be a very potent negotiating tactic which could lead to the first-best solution, namely stringent commitments to impose a carbon price and allow the full pass-through of carbon costs by developing countries.

REFERENCES

AEA Technology Environment and Ecofys UK (2006), *LETS Update: Decision Makers Summary*, Bristol: Environment Agency, April.

Ahner, Nicole (2009), "Final Instance: World Trade Organization - Unilateral Trade Measures in EU Climate Change Legislation", EUI Working Paper RSCAS 2009/58, European University Institute, Robert Schuman Centre for Advanced Studies and Loyola De Palacio Programme on Energy Policy, Florence.

Amann, Markus, Janusz Cofala, Peter Rafaj and Fabian Wagner (2009), *The Impact of the Economic Crisis on GHG Mitigation Potentials and Costs in Annex I Countries*, GAINS report, International Institute for Applied Systems Analysis (IIASA), Laxenburg, November.

Baer, P., T. Athanasiou and S. Kartha (2007), *The Right to Development in a Carbon Constrained World – The Greenhouse Development Rights Framework*, Heinrich Böll Foundation, Berlin.

Behrens, A. (2008), *Financial Impacts of Climate Change: An Overview of Climate Change related Actions in the European Commission's Development Cooperation*, CEPS Working Document No. 305, Centre for European Policy Studies, Brussels.

Behrens, Arno, Jorge Núñez Ferrer and Christian Egenhofer (2008), *Financial Impacts of Climate Change: Implications for the EU Budget*, CEPS Working Document No. 300, Centre for European Policy Studies, Brussels, August.

Biermann, F. and R. Brohm (2005), "Implementing the Kyoto Protocol without USA: The strategic role of energy tax adjustments at the border", *Climate Policy* 4, pp. 289-302.

Bin, S. and Harriss, R. (2008), *Talking Carbon: Implications of US-China Trade*, ICTSD Selected Issue Brief No. 2, International Centre for Trade and Sustainable Development, Geneva.

Bordoff, J. (2008), "International Trade Law and the Economics of Climate Policy: Evaluating the Legality and Effectiveness of Proposals to Address Competitiveness and Leakage Concerns", *Brookings Global Economy and Development*, Brookings Institution, Washington, D.C.

China Sustainable Energy Program (2008), "Fact Sheet: China Emerging as New Leader in Clean Energy Policies" (http://www.efchina.org).

Council of the EU (2003), Council Directive 2003/96/EC of 27 October 2003 restructuring the Community framework for the taxation of energy products and electricity (Text with EEA relevance), *Official Journal L 283*, 31 October, pp. 51-70.

Council of the EU (2007), New Impetus for EU Environmental Policy – Draft Council conclusions, 10796/07 (available at www.consilium.europa.eu).

Demailly, D. and P. Quirion (2005), "Leakage from climate policies and border-tax adjustment: Lessons from a geographic model of the cement industry", CESifo Venice Summer Institute.

den Elzen, M.G.J. and N. Höhne (2008), "Reductions of greenhouse gas emissions in Annex I and non-Annex I countries for meeting concentration stabilisation targets", *Climatic Change*, No. 91, pp. 249–274.

Diringer, Elliot (2008), *The U.S. Election and Prospects for a New Climate Agreement*, Transatlantic Climate Policy Group, Heinrich-Böll-Stiftung, Washington, D.C., 22 October.

Egenhofer, Christian (2007), "The Making of the EU Emissions Trading Scheme: Status, Prospects and Implications for Business", *European Management Journal*, Vol. 25, No. 6, December, pp. 453-463.

Egenhofer, Christian (2009), "The New EU Emissions Trading Scheme: A Blueprint for the Global Carbon Market?", in J. Harris and N. Goodwin (eds), *21st Century Macroeconomics: Responding to the Climate Challenge*, Cheltenham UK and Northampton MA: Edward Elgar Publishing, Chapter 11, pp. 246-263.

Egenhofer, Christian, Arno Behrens and Anton Georgiev (forthcoming), "EU Strategies for Climate Change Policy beyond 2012", in Hans Günter Brauch; Úrsula Oswald Spring, John Grin, Czeslaw Mesjasz, Patricia Kameri-Mbote, Navnita Chadha Behera, Béchir Chourou and Heinz Krummenacher (eds), *Facing Global Environmental Change - Environmental, Human, Energy, Food, Health and Water Security Concepts*, Berlin: Springer Hexagon Series on Human and Environmental Security and Peace, Volume 4.

Ellerman, A. Denny and Paul L. Joskow (2008), *The European Union's Emissions Trading System in perspective*, Pew Center on Global Climate Change, Washington, D.C., May.

Ellerman, A. Denny, Barbara K. Buchner and Carlo Carraro (eds) (2007), *Allocation in the European Emissions Trading Scheme. Rights, Rents and Fairness*, Cambridge: Cambridge University Press.

Ellerman, A. Denny, Frank J. Convery and Christian De Perthuis with contributions by Emilie Alberola, Barbara K. Buchner, Anaïs Delbosc, Cate Hight, Jan Horst Keppler and Felix C. Matthes (2010), *Pricing Carbon: The European Union Emissions Trading Scheme*, Cambridge: Cambridge University Press, January.

Energy Information Administration (EIA) (2009), *International energy outlook 2005*, Washington, D.C.

Environment Canada (2009), *Canada's 2007 Greenhouse Gas Inventory – A Summary of Trends*, Greenhouse Gas Division (http://www.ec.gc.ca/pdb/ghg/inventory_report/2007/som-sum_eng.cfm).

European Commission (2007), *A European strategic energy technology plan (SET Plan) - Towards a low carbon future*, Communication from the Commission to the Council, the European Parliament, the European Economic and Social Committee and the Committee of the Regions of 22 November 2007, COM(2007)723 final, Brussels, November.

European Commission (2008a), *Impact Assessment – Document accompanying the Package of Implementation measures for the EU's objectives on climate change and renewable energy for 2020*, Commission Staff Working Document SEC(2008)85, Brussels, January.

European Commission (2008b), *Annex to the Impact Assessment – Document accompanying the Package of Implementation measures for the EU's objectives on climate change and renewable energy for 2020*, Commission Staff Working Document, SEC(2008)85 VOL. II, Brussels, February.

European Commission (2009a), *Towards a comprehensive climate change agreement in Copenhagen*, Communication from the Commission to the Council, the European Parliament, the European Economic and Social Committee and the Committee of the Regions, COM(2009)39 final, Brussels, January.

European Commission (2009b), "Additional Background Information", Commission Staff Working Document accompanying the Communication from the Commission to the European Parliament, the Council, the European Economic and Social Committee and the Committee of the Regions: *Towards a comprehensive climate change agreement in Copenhagen*, SEC(2009)101, Brussels, January.

European Commission (2009c), "Extensive Background Information and Analysis – Part 1", Commission Staff Working Document accompanying the Communication from the Commission to the European Parliament, the Council, the European Economic and Social Committee and the Committee of the Regions: *Towards a comprehensive climate change agreement in Copenhagen*, SEC(2009)102, Brussels, January.

European Commission (2009d), "Extensive Background Information and Analysis – Part 2", Commission Staff Working Document accompanying the Communication from the Commission to the European Parliament, the Council, the European Economic and Social Committee and the Committee of the Regions: *Towards a comprehensive climate change agreement in Copenhagen*, SEC(2009)102, Brussels, January.

European Commission (2009e), *Investing in the Development of Low Carbon Technologies (SET-Plan)*, Communication from the Commission to the European Parliament, the Council, the European Economic and Social Committee and the Committee of the Regions, COM(2009)519 final of 7.10.2009.

European Commission (2010a), "Commission Decision of 24 December 2009 determining, pursuant to Directive 2003/87/EC of the European Parliament and of the Council, a list of sectors and subsectors which are deemed to be exposed to a significant risk of carbon leakage (notified under document C(2009) 10251)", *Official Journal of the European Union*, L 001, 05/01/2010, pp. 0010 - 0018, European Union, Brussels, January.

European Commission (2010b), *International climate policy post-Copenhagen: Acting now to reinvigorate global action on climate*, Communication from the Commission to the Council, the European Parliament, the European Economic and Social Committee and the Committee of the Regions, COM(2010)86 final, Brussels, March.

European Council (2007), *Brussels European Council 8/9 March 2007 - Presidency Conclusions*, 72241/1/07 REV 1 CONCL 1, Council of the European Union, Brussels, May.

European Council (2008), *Brussels European Council 19/20 June 2008 - Presidency Conclusions*, 11018/1/08 REV 1 CONCL 2, Council of the European Union, Brussels, June.

European Council (2009), *Brussels European Council 18/19 June 2009 - Presidency Conclusions*, 11225/09 CONCL 2, Council of the European Union, Brussels, June.

European Environment Agency (EEA) (2007), *EU Energy in Figures*, Pocket Book 2007, December.

European Environment Agency (EEA) (2009), *Annual European Community greenhouse gas inventory 1990–2007 and inventory report 2009: Submission to the UNFCCC Secretariat*, EEA Technical Report No. 4/2009, Copenhagen.

European Parliament (2007), Resolution on Trade and Climate Change, 29 November (P6_TA(2007)0576).

European Parliament (2008), European Parliament resolution of 24 April 2008 on the Green Paper on market-based instruments for environment and related policy purposes (2007/2203(INI)) (www.europarl.europa.eu).

European Parliament and Council of the European Union (2009a), "Directive 2009/29/EC of the European Parliament and of the Council of 23 April 2009 amending Directive 2003/87/EC so as to improve and extend the greenhouse gas emissions allowance trading scheme of the Community", *Official Journal of the European Union*, OJ L 140/63, 5 June 2009, European Union, Brussels, June, pp. 63-87.

European Parliament and Council of the European Union (2009b), "Decision No 406/2009/EC of the European Parliament and of the Council of 23 April 2009 on the effort of Member States to reduce their greenhouse gas emissions to meet the Community's greenhouse gas emissions reduction commitments up to 2020", *Official Journal of the European Union*, OJ L 140/136, 5 June, European Union, Brussels, June, pp. 136-148.

European Parliament and Council of the European Union (2009c), "Directive 2009/28/EC of the European Parliament and of the Council of 23 April 2009 on the promotion of the use of energy from renewable sources and amending and subsequently repealing Directives 2001/77/EC and 2003/30/EC", *Official Journal of the*

European Union, OJ L 140/16, 5 June, European Union, Brussels, June, pp. 16-62.

Frankel, Jeffrey A. (2009), "Addressing the Leakage/Competitiveness Issue", in Isaac Sorkin and Lael Brainard (eds), *Climate Change Policy Proposals. Climate Change, Trade, and Competitiveness: Is a Collision Inevitable?*, Brookings Institution Press, Washington, D.C.

Fujiwara, Noriko and Christian Egenhofer (2010), Understanding India's Climate Agenda, CEPS Policy Brief No. 206, Centre for European Policy Studies, Brussels, February.

GATT (General Agreement on Tariffs and Trade) (1970), Working Party Report on Border Tax Adjustments, GATT Document L/3464 (http://www.worldtradelaw.net/reports/gattpanels/bordertax.pdf).

Genasci, M. (2008), "Border Tax Adjustments and Emissions Trading: The Implications of International Trade Law for Policy Design", *Carbon and Climate Law Review*, Issue 01/2008.

Gros, Daniel (2009), *Global Welfare Implications of Carbon Border Taxes*, CEPS Working Document No. 315, Centre for European Policy Studies, Brussels, July.

Grossman, G.M. (1980), "Border Tax Adjustments: Do they distort trade?" *Journal of International Economics* 10, pp. 117-128.

Gupta, S., D.A. Tirpak, N. Burger, J. Gupta, N. Höhne, A.I. Boncheva, G.M. Kanoan, C. Kolstad, J.A. Kruger, A. Michaelowa, S. Murase, J. Pershing, T. Saijo and A. Sari (2007), "Policies, Instruments and Co-operative Arrangements", in B. Metz, O.R. Davidson, P.R. Bosch, R. Dave, L.A. Meyer (eds), *Climate Change 2007: Mitigation*, contribution of Working Group III to the Fourth Assessment Report of the Intergovernmental Panel on Climate Change, Cambridge and New York: Cambridge University Press, pp. 745-807.

Gurria, Angel (2009) "Carbon has no place in global trade rules", *Financial Times*, 4 November.

Hoerner, J.A. (1998), "The Role of Border Tax Adjustments in Environmental Taxation: Theory and US Experience", International Workshop on Market-Based Instruments and International Trade of the Institute for Environmental Studies, Amsterdam, conference proceedings.

Houser, Trevor, Rob Bradley, Britt Childs, Jacob Werksman and Robert Heilmayr (2008), *Leveling the Carbon Playing Field: International*

Competition and US Climate Policy Design, Peterson Institute for International Economics and World Resources Institute, Washington, D.C.

Houser, Trevor, Rob Bradley, Britt Childs, Jacob Werksman and Robert Heilmayr (2008), *Leveling the Carbon Playing Field: International Competition and US Climate Policy Design*, Peterson Institute for International Economics and World Resources Institute, Washington, D.C.

Howard, Steve and Changhua Wu (2008), *China's Clean Revolution*, The Climate Group (http://www.theclimategroup.org).

Hufbauer, Gary C., Steve Charnovitz and Jisun Kim (2009), *Global Warming and the World Trade System*, Petersen Institute for International Economics, Washington, D.C.

IEA (International Energy Agency) (2007), *World Energy Outlook 2007: China and India Insights*, OECD/IEA, Paris.

IEA (International Energy Agency) (2008), *Energy Technology Perspectives 2008*, OECD/IEA, Paris.

IEA (International Energy Agency) (2009a), *CO_2 Emissions from Fuel Combustion, 2009 Edition*, OECD/IEA, Paris.

IEA (International Energy Agency) (2009b), *World Energy Outlook 2009*, OECD/IEA, Paris.

IEA/OECD (2009), "Energy, Carbon and Sulphur Dioxide Taxation", in *Addressing Climate Change – Policies and Measures* (online database) (http://www.iea.org/textbase/pm/index_clim.html).

IISD (International Institute for Sustainable Development) (2009a), "Summary of the Bonn climate change talks", *Earth Negotiations Bulletin*, Vol. 12, No. 427, Winnipeg, Manitoba, Canada, 17 August.

IISD (International Institute for Sustainable Development) (2009b), "Summary of the Bangkok climate change talks', *Earth Negotiations Bulletin*, Vol. 12, No. 439, Winnipeg, Manitoba, Canada, 12 October.

IMF (International Monetary Fund) (2009), *World Economic Outlook Database*, October.

IPCC (2007), *Climate Change 2007: Synthesis Report*, Summary for Policy Makers, Intergovernmental Panel of Climate Change, Geneva.

Ismer, R. and K. Neuhoff (2004), *Border Tax Adjustments: A Feasible Way to Address Non-Participation in Emissions Trading*, Cambridge Working Papers in Economics, CWPE 0409.

Johnson, H.G. and M. Krauss (1970), "Border Taxes, Border Tax Adjustment, Comparative Advantage and the Balance of Payments", *Canadian Journal of Economics*, November.

Kartha, S, B. Kjellén, P. Baer and T. Athanasiou (2008), "Linking measurable, reportable and verifiable mitigation actions by developing countries to measurable, reportable and verifiable financial and technical support by developed countries", in C. Egenhofer et al., *Positive Incentives*, ECP Report No. 5, European Climate Platform, June.

Kettner, Claudia, Angela Köppl, Stefan P. Schleicher and Gregor Thenius (2007), *Stringency and Distribution in the EU Emissions Trading Scheme – the 2005 Evidence*, Nota di Lavora 22.2007, Fondazione Eni Enrico Mattei, Milan.

Levin, Kelly and Rob Bradley (2010), *Comparability of Annex I Emission Reduction Pledges*, WRI Working Paper, World Resources Institute, Washington, D.C., February.

Lockwood, B. and J. Whalley (2008), *Carbon Motivated Border Tax Adjustments: Old Wine in Green Bottles?*, NBER Working Paper No. 14025, National Bureau of Economic Research, Cambridge, MA.

Majocchi, A. and M. Missaglia (2002), *Environmental taxes and border tax adjustment*, Working Paper No. 127, Societa Italiana Economia pubblica (SIEP), Turin.

Martinez, Kim K. and Karsten Neuhoff (2005), "Allocation of carbon emissions certificates in the power sector: how generators profit from grandfathered rights", *Climate Policy*, London: Earthscan, Vol. 5, No. 1, September, pp. 61-78.

Mathiesen, L. and O. Maestad (2002), "Climate policy and steel industry: Achieving global emissions reductions by an incomplete climate agreement", Discussion Papers 20/02, Norwegian School of Economics and Business Administration, Bergen.

Matthes, Felix C. and Karsten Neuhoff (2007), *Auctioning in the European Union Emissions Trading Scheme,* Report commissioned by WWF, Berlin/Cambridge: Öko-Institut - University of Cambridge, September.

Matthes, Felix C., Verna Graichen and Jack Repenning (2005), *The environmental effectiveness and economic efficiency of the European Union Emissions Trading Scheme: Structural aspects of allocation*, Report to the WWF, Freiburg: Öko-Institut.

McKibben, W.J. and P. Wilcoxen (2008), "The Economic and Environmental Effects of Border Tax Adjustments for Climate Policy", Brookings Global Economy and Development Conference, Brookings Institution, Washington, D.C.

McKinsey & Company (2009), *Pathways to a Low Carbon Economy*, Version 2 of the Global GHG Abatement Cost Curve.

Mission of the People's Republic of China to the European Union (2009), "China Announces Targets on Carbon Dioxide Emissions Cuts", Newsletter, Issue No. 26, Mission of the People's Republic of China to the European Union, Brussels, November.

NDRC (National Development and Reform Commission) (2005), *National Energy Plan*, Chinese government document.

Neufeldt, Henry, Detlef P. van Vuuren, Morna Isaac, Brigitte Knopf, Ottmar Edenhofer, Wolfgang Schade, Eberhard Jochem and Frans Berkhout (2009), *Reaching the 2°C Target: Technological Requirements, Economic Costs and Policies*, CEPS/ADAM Policy Brief, Centre for European Policy Studies, Brussels, May.

Oxfam (2007), *Adapting to climate change – What's needed in poor countries, and who should pay*, Oxfam Briefing Paper 104, Oxfam, Washington, D.C.

Pauwelyn, Joost (2007), "U.S. Federal Climate Policy and Competitiveness Concerns: The Limits and Options of International Trade Law", Working Paper 07-02, Nicholas Institute for Environmental Policy Solutions, Duke University, Durham, N.C., April.

Peters, Glen and Edgar Hertwich (2008), "CO_2 embodied in international trade with implications for global climate policy", *Environmental Science and Technology*, Vol. 42, No. 5, pp. 1401-1407.

Quick, Reinhard (2008), "Border Tax Adjustment in the Context of Emissions Trading: Climate Protection or Naked Protectionism", *Global Trade and Customs Journal*, Kluwer Law International.

Quick, Reinhard (2009), "Border Tax Adjustment to Combat Carbon Leakage: A Myth", *Global Trade and Customs Journal*, Vol. 4, No. 11/12, Kluwer Law International.

Radunsky, Klaus and Claudia Laabs (2009), "ISO's contribution to the management of GHG emissions in the global supply chain", *ISO Focus Magazine*, 15 February.

Reinfeldt, Fredrik (2009), "Taking on the challenge - The Swedish Presidency of the European Union", speech at CEPS, Brussels, 9 June (http://www.regeringen.se/sb/d/11738/a/128010).

Renaud, Julia (2008), "Issues Behind Competitiveness and Carbon Leakage, Focus on Heavy Industry", International Energy Agency, October (http://www.iea.org/textbase/papers/2008/Competitiveness_and_Carbon_Leakage.pdf).

Sampson, Gary P. (1999), "WTO Rules and Climate Change: The Need for Policy Coherence", in W. Bradnee Chambers (ed.), *Global Climate Governance*, Tokyo: United Nations University Press.

Sarkozy, Nicolas (2009), Discours de M. le Président de la République sur la stratégie de la France dans la lutte contre le réchauffement climatique, September (http://www.elysee.fr/documents/index.php?mode=cview&cat_id=7&press_id=2898&lang=fr).

Schmidt, R.C. and R. Marschinski (2009), "Can China Benefit from Adopting a Binding Emissions Target?", Working Paper, December (http://lehre.wiwi.hu-berlin.de/Professuren/vwl/wp/mitarbeiter/papert/China_paper).

Seligsohn, Deborah and Hilary McMahon (2009), *Measuring Climate Change Progress in China*, World Resources Institute, February (http://www.wri.org).

Sinn, Hans-Werner (2008), "Public policies against global warming: A supply side approach", *Journal of International Tax and Public Finance*, Vol. 15, No. 4, August.

Spencer Thomas, Kristinan Tangen and Anna Korppoo (2010), "The EU and the Global Climate Regime, Getting Back in the Game", Briefing Paper 55, Finnish Institute of International Affairs (UPI), Helsinki, 25 February.

Swedish Energy Agency (2006), *The EU Emissions Trading Scheme after* 2012, Swedish Energy Agency and the Swedish Environmental Protection Agency, Stockholm.

Tirole, Jean (2009a), "Politique climatique: une nouvelle architecture internationale", Conseil d'Analyse économique, October (http://www.voxeu.org/index.php?q=node/4210).

Tirole, Jean (2009b), "Climate change negotiations: Time to reconsider", VoxEU, 16 November (http://www.voxeu.org/index.php?q=node/4210.)

UNDP (United Nations Development Programme) (2009), *Human Development Report 2009,* United Nations, New York (http://hdrstats.undp.org/en/indicators/91.html).

UNFCCC (2008), *Greenhouse Gas Inventory Data* (http://unfccc.int/ghg_data/ghg_data_unfccc/items/4146.php).

Van Asselt, H. and F. Biermann (2007), "European emissions trading and the international competitiveness of energy-intensive industries: A legal and political evaluation of possible supporting measures", *Energy Policy* 35, pp. 497-506.

Veenendaal, P. and T. Manders (2008), *Border tax adjustment and the EU-ETS, a quantitative assessment,* CPB Document No. 171, Central Planning Bureau, The Hague.

Weber, Christopher L., Glen Peters, Dabo Guan and Klaus Hubacek (2008), "The contribution of Chinese exports to climate change", *Energy Policy,* Vol. 36, No. 9, pp. 3572-3577.

Werksman, J.D. and T.G. Houser (2008), "Competitiveness, leakage and comparability: Disciplining the use of trade measures under a post-2012 climate agreement", Discussion Paper, World Resources Institute, December.

World Bank (2008), "International Trade and Climate Change: Economic, Legal and Institutional Perspectives", Washington, D.C.

WTO/UNEP (2009), *Trade and Climate Change,* World Trade Organisation, Geneva.

Yamin, F. and J. Depledge (2004), *The international climate change regime: a guide to rules, institutions and procedures,* Cambridge: Cambridge University Press.

Yan, Xu and Kejun Jiang (2009), "China's Energy and Climate Policies", presentation at the ESRI International Collaboration Projects 2008, Tokyo Meeting, 23 February.

Zetterberg, L., K. Nilsson, M. Åhman, A.-S. Kumlin and L. Birgersdotter (2004), *Analysis of National Allocation plans for the EU ETS,* IVL Report B1591, Swedish Environmental Research Institute, Stockholm.